Henry B. Parsons.

Pharmacy Class of '76.

Ann Arbor.

Mich.

Nos. 1-2. Tartaric & Citric acids
" 3-4- Tannic & Gallic "
" 10-12 - Benzoic & Salicylic "
" 17-18- Butyric & Valeric. "
" 13-32 - Phenol & Creasote
" 58-59- Methyl & Ethyl Alcohols.
" 60 - 59 - Amyl & " "
" 63 - 59 - Chloroform & " alcohol.
" 64 - 25 - Chloral Hydrate & Glycerine
" 33 - 35 - Benzene & Petroleum Naphtha
" 25 - 55 - Soap & Glucose.
" 56 - 53 - Dextrine & Sucrose.
" 52 - 55 - Acacia & Glucose.
" 51 - 50 - Gelatin & albumen.
" 22 - 30 - Turpentine & Castor Oils
" 22 - 20 - Castor & Olive oils.
" 21 - 23 - Linseed & Cottonseed Oils.
" 37- 40 - Quinia & Quinidia.
" 42-38 - Morphia & Cinchonia.
" 46 + 49 - Caffeine & Salicine.

OUTLINES

OF

PROXIMATE ORGANIC ANALYSIS.

OUTLINES

OF

PROXIMATE ORGANIC ANALYSIS.

FOR THE

IDENTIFICATION, SEPARATION, AND QUANTITATIVE DETERMINATION

OF THE

MORE COMMONLY OCCURRING ORGANIC COMPOUNDS.

BY

ALBERT B. PRESCOTT,

PROFESSOR OF ORGANIC AND APPLIED CHEMISTRY IN THE UNIVERSITY OF MICHIGAN.

NEW YORK:
D. VAN NOSTRAND, PUBLISHER,
23 MURRAY STREET, AND 27 WARREN STREET.

1875.

PREFACE.

THIS little work has been prepared more especially for the use of a class of chemical students who devote a semester to the analysis of vegetable products and other organic mixtures, taking previously at least two semesters in qualitative and quantitative analysis. After working with this class for several years, without other aid than a manuscript digest of directions and references, the author is convinced that a compilation in this subject is desirable—not alone for students in special applications of chemistry, but for the convenience of every general analyst.

Proximate organic analysis is not altogether impracticable, and organic chemistry is not solely a science of synthetical operations even at present. It is true, as the chief analytical chemists have repeatedly pointed out, that in the rapid accumulation of organic compounds the means of their identification and separation have been left in comparative neglect. It is true, also, that the field is limitless; but this is not a reason for doing nothing in it. Fifty years ago, the workers in inorganic analysis were unprovided with a comprehensive system, but they went on exploring the mineral kingdom and using their scanty means to gain valuable results.

That this compilation is a fragmentary and very brief exponent of this part of analytical science as it exists at present, the author is fully aware, but he hopes that, as a beginning, it may prove to be worth enough to afford an opportunity for its improvement hereafter.

University of Michigan, September, 1874.

CONTENTS.

OUTLINES

OF

PROXIMATE ORGANIC ANALYSIS.

PRELIMINARY EXAMINATIONS.

1. CARBON (uncombined) is recognized by its sensible properties (as charcoal, graphite, or diamond), by not vaporizing when heated, and by resisting ordinary solvents—neutral, alkaline or acid—except that graphite is oxidized by digestion with chlorates and sulphuric or hydrochloric acid, or with bichromates and sulphuric acid, or with mixed nitric and sulphuric acids.—Also, on ignition in the air, or in a close tube with oxide of copper, carbonic anhydride is obtained from carbon alone, as well as from its compounds.

2. THE COMPOUNDS OF CARBON—except the alkaline carbonates—yield carbonic anhydride when ignited in the air or in a tube with supply of oxygen (as with dry oxide of copper). The non-volatile "*Organic*" *Compounds of Carbon* leave a residue of carbon after partial combustion—*i.e.*, they carbonize by ignition.

3. Preliminary examination OF SOLIDS to determine whether inorganic or organic, or both.

a. Heat gradually, to prolonged ignition, in a glass tube open at both ends, or on platinum foil.

(1) The substance is permanent: Inorganic.

(2) Carbonizes and burns away, leaving no residue: Organic. See 5, *a*.

(3) Carbonizes and leaves a fixed residue: Organic and Inorganic. See *c*.

(4) There is doubt as to carbonization: test according to *b*.

(5) The substance vaporizes—wholly or partly: test according to *b*. Also consider ammonium salts, the volatile elements, and the inorganic volatile acids, oxides, sulphides, etc. Examine according to 4, *b*.

b. Mix the (dry) substance (free from carbonates yielding CO_2 on ignition) with dry oxide of copper; introduce into a short combustion-tube or a hard-glass test-tube; connect, by a cork and bent narrow tube, with a solution of lime or baryta, or basic acetate of lead, and ignite. If a precipitate is formed, test it as a carbonate.

c. Ignite a portion in a porcelain capsule, until free from carbon—cooling and adding a drop or two of concentrated nitric acid from time to time, if necessary to facilitate the combustion. Submit the residue to inorganic analysis. Examine another portion for organic bodies—applying the solvents, as in 134 (9) or (7). For an index of some of the most common organic solids, see 5, *a*.

4. Preliminary examination of LIQUIDS, to determine whether partly or wholly organic or not, and to separate dissolved solids.

a. Evaporate a portion, on a slip of glass, at a very gentle heat. If, after cooling, a solid residue is obtained, test it according to 3. If there is an insufficient residue, obtain for this examination a larger quantity by distillation, as directed in *b*.

b. Distil from a small retort or connected flask, admitting a thermometer, using a very gradually-increasing heat, and changing the receiver as often as the boiling point is seen to rise. Cool the residue and distillates. Test the solid portions according to 3; the liquid portions, also, according to 3, *a* or *b*—then referring as indicated in the next paragraph. For index of Organic Liquids, see 5, *b*.

5. *a.* SOLIDS.

NON-VOLATILE.

Acids: Aconitic—9.
Boheic—21.
Caffetannic—20.
Catechuic—18.
Catechutannic—17.
Carminic—26.
(Chrysophanic)—27.
Citric—8.
Columbic—24.
Digitalic—12.
(Gallic)—14.
Gambogic—28.
Gentianic—25.
Malic—10.
Meconic—11.
Morintannic—19.
(Pyrogallic)—15.
Quinic—22.
Quinotannic—16.
Quinovic—23.
Racemic—7.
Tannic—13.
Tartaric—6.
Santalic—29.

Fatty Acids:
Cerotic—54.
Erucic (melts at 34° C.)—49.
Lauric—50.
Myristic—51.
Palmitic—52.
Stearic—53.

Fixed salts of volatile acids.

Fixed Oils—59, *b*; 60 to 63.
Soaps—67.
Resins—99, and 68 to 98.
Alkaloids (fixed)—132 to 143.
Carbohydrates:
Cellulose—185.
Dextrin—175.
Gum—172.
Gun-cotton—186.
Pectin, etc.—177 to 184.
Starch—176.
Sugars—187 to 190.
Albumenoids—162 to 167.
Gelatin—170, 171.

VOLATILE.

Acids: Benzoic—30.
(Chrysophanic)—27.
Cinnamic—31.
(Gallic)—14.
Nitrophenic—36.
(Pyrogallic)—15.
Salicylic—33.
Succinic—32.
Sulphophenic—37.
Veratric—34.

Camphors—115, 101, and 111.
Anthracene—117.
Alizarin—118.
Anilin compounds—123.
Chloral hydrate—198.
Iodoform—199.
Salts of Volatile Alkaloids.

b. LIQUIDS.

NON-VOLATILE.

Acid: Lactic—38.
Fatty Acids:
Linoleic (melts, 18° C.)—48.
Oleic—47.
Ricinoleic—46.
Fixed Oils—59.
(Soft Soaps)—67.
Glycerin—66.

VOLATILE.

Acids: Acetic—40.
Butyric—41.
Formic—39.
Valeric—42.
Volatile Oils—105, 104, and 100 to 114.
Creosote—116.
Volatile Alkaloids—131 and 126 to 130.

Anilin—121.
Solvents:
Alcohol—192.
Aldehyd—193.
Amyl. Alcohol—201.
Benzole—119.
Chloroform—197.

Solvents—Continued.
Ether—195.
Co. Ethers—40*b*, 41*b*, 42*a*, 44, etc.
Meth. Alcohol—191.
Nitrobenzole—120.
Petroleum—119½.

SOLID NON-VOLATILE ACIDS.

6. TARTARIC ACID. $H_2C_4H_4O_6$. *Characterized* by the form of its crystals and its rotation of polarized light (*a*); by its odor when heated, and its color when treated with sulphuric acid (*b*); by the properties of its salts of calcium, potassium, lead, and silver (*c*); by the extent of its reducing power (*d*).—*Separated* (as free acid) from salts or other substances insoluble in alcohol by its solubility in that menstruum, and from aqueous solutions by its solubility in amylic alcohol (*e*); from alcoholic solutions by the insolubility of tartrates in alcohol (*c*); from citric acid by the precipitation of calcium tartrate in cold water and of potassium tartrate in aqueous alcohol (*c*); from substances not precipitable by oxide of lead by the method given under Acetic acid at *g* (40).—*Determined* by acidimetry (*f*); gravimetrically as lead, calcium, or potassium tartrate (*g*); by sp. gr. of water solutions (see Storer's "Dictionary of Solubilities").

a. Ordinary tartaric acid, or "dextrotartaric acid," *crystallizes* in colorless, transparent, hard, monoclinic (oblique rhombic) prisms, permanent in the air, *soluble* in 1.5 parts cold water, 0.5 part hot water, 3 parts alcohol, not soluble in ether. The solution rotates the plane of polarized light to the right.

b. When *heated* to 170° to 180° C., the crystals melt with formation of metatartaric acid, etc.; by higher heat in the air, various distillation products are generated, and the mass burns with the *odor* of burnt sugar and the separation of carbon.—

Pure tartaric acid dissolves in cold concentrated **sulphuric acid,** colorless, the solution turning black when warmed.

c. The normal tartrates of potassium, sodium, and ammonium, and the acid tartrate of sodium, are freely *soluble* in water; the acid tartrates of potassium and ammonium are sparingly soluble in water; the normal tartrates of non-alkaline metals are insoluble or only slightly soluble in water, but mostly dissolve in solution of tartaric acid. Tartrates are insoluble in absolute alcohol. Aqueous alkalies dissolve most of the tartrates (those of mercury, silver, and bismuth being excepted), generally by formation of soluble double tartrates. For this reason, tartaric acid prevents the precipitation of salts of iron and many other heavy oxides by alkalies. Hydrochloric, nitric, and sulphuric acids decompose tartrates.

A solution of tartaric acid added to cold solution of **lime,** leaving the reaction alkaline, causes a slight white precipitate of calcic tartrate (distinction from Citric acid, which precipitates only when heated). The same precipitate is produced with a tartrate and calcic chloride solution; but not readily, if at all, with free tartaric acid and calcic sulphate solution (distinction from Racemic acid). The precipitate of calcic tartrate is soluble in cold solution of potassa, is precipitated gelatinous on boiling, and again dissolves on cooling (distinctions from Citrate), and is dissolved by acetic acid (distinction from Oxalate).

Solution of **potassa,** or potassic acetate, precipitates concentrated solutions of tartaric acid, as the acid tartrate of potassium in microscopic crystals of the trimetric system, soluble in alkalies and in mineral acids, not soluble by acetic acid. The precipitate is soluble in 230 parts of water at 15°, or in 15 to 20 parts of boiling water, but insoluble in **alcohol,** the addition of which promotes its formation in water solutions (distinction and separation from Citric, Oxalic, and Malic acids).—Tartaric acid is distinguished from citric acid, in crystal, and the former is detected in a crystalline mixture of the two acids, as follows: *

* Hager's "Untersuchungen," B. 2, S. 103.

A solution of 4 grammes of dried potassa in 60 cubic centimeters of water and 30 cubic centimeters of 90 per cent. alcohol is poured upon a glass plate or beaker-bottom to the depth of about 0.6 centimeter (one-fourth inch). Crystals of the acid under examination are placed, in regular order, three to five centimeters (one to two inches) apart, in this liquid, and left without agitation for two or three hours. The citric acid crystal dissolves slowly but completely and without losing its transparency. The tartaric acid crystal (or the crystal containing tartaric acid) becomes, in a few minutes, opaque white (in a greater or less degree), and continues for hours and days slowly to disintegrate without dissolving and with gradual projection of spicate *crystals,* fibrous and opaque, also trimetric prisms. (See, also, Citric acid, *e.*)

Solution of **lead** acetate precipitates free tartaric acid or tartrates as white normal tartrate of lead, very slightly soluble in water, insoluble in alcohol, but slightly soluble in acetic acid, readily soluble in tartaric acid and in tartrate of ammonium solution, and freely soluble in ammoniacal solution of tartrate of ammonium (distinction from Malate), somewhat soluble in chloride of ammonium.

Solution of **silver** nitrate precipitates solutions of normal tartrates (not free tartaric acid) as white argentic tartrate, soluble in ammonia and in nitric acid. On boiling, the precipitate turns black, by reduction of silver, some portion of which usually deposits as a mirror-coating on the glass. The mirror is formed more perfectly if the washed precipitate of argentic tartrate is treated with ammonia just enough to dissolve nearly all of it, and the solution left on the water bath. (The reduction is a distinction from Citrate). Free tartaric acid does not reduce silver from its nitrate.

d. The **copper** sulphate with potassa is not reduced by tartaric acid. Potassium **permanganate** solution is reduced very slowly by free tartaric acid; but quickly by alkaline solution of tartrates, with separation of brown binoxide of manganese (distinction from Citrates which separate the brown binoxide of manganese slowly or not at all, leaving green solution of manganate).

e. Tartaric acid may be extracted from tartrates by decomposing with sulphuric acid and dissolving with **alcohol**, sulphates being generally insoluble in alcohol. Free tartaric acid may be extracted from water solutions by agitation with **amylic alcohol**, which rises to the surface.

Quantitative.—*f.* Free tartaric acid, unmixed with other acids, may be determined *volumetrically* by adding a normal solution of soda, to the neutral tint of litmus. Weighing 7.500 grammes, the required number of cubic centimeters of normal solution equals the number per cent. of acid.

g. In absence of acids forming insoluble lead salts, tartaric acid may be precipitated by acetate of lead solution, washed with dilute alcohol, dried on the water bath and weighed as normal lead tartrate. $\mathbf{PbC_4H_4O_6 : H_2C_4H_4O_6}$: : 1 : 0.422535.

In absence of acids forming insoluble calcium salts, tartaric acid may be precipitated from solution of neutral sodium tartrate by chloride of **calcium.** If ammonium salts are present, the ammonia should first be mostly expelled by adding sodium carbonate and heating—the excess of carbonate being neutralized with acetic acid. The precipitate of calcium tartrate should be heated and left aside for completion, washed with a little water and then with dilute alcohol, and dried (in a tared filter) at 40° to 50° C. $\mathbf{Ca\,C_4H_4O_6{+}4H_2O : H_2C_4H_4O_6}$: : 1 : 0.577.

In presence of citric acid, oxalic acid, sulphuric acid, phosphoric acid, etc., the tartaric acid may be determined as **potassium** bitartrate. The solution of acid is made nearly neutral by addition of soda, or the solution of salt (tartrate) is made slightly acid by addition of acetic acid; this water solution is obtained in concentrated form and treated with a little alcohol but not to cause a precipitate, and then precipitated with concentrated solution of acetate of potassium. The precipitate is washed with alcohol, and dried on the water bath. $\mathbf{KH\,C_4H_4O_6 : H_2C_4H_4O_6}$: : 1 : 0.797. Results approximate.

7. RACEMIC ACID. Isomer of tartaric acid, from which it is *distinguished* as follows: By forming triclinic *crystals*, $H_2C_4H_4O_6 . H_2O$; *soluble* in 5 parts cold water or 48 parts of alcohol of sp. gr. .809; slightly efflorescent on the surface; losing the water of crystallization at 100°. By its solution (uncombined) being able to form after a short time a slight precipitate in solution of calcic sulphate and a precipitate in solution of calcic chloride; the precipitate of calcic racemate being, after solution in hydrochloric acid, precipitated again by ammonia, that is, not soluble in chloride of ammonium solution. By being inactive toward polarized light.

8. CITRIC ACID. $H_3C_6H_5O_7$. *Characterized* by the form, solubilities, and fusibility of its crystals (*a*); by the properties of its salts of calcium, barium, lead, silver, potassium (*b*); by the limits of its reducing power (*c*).—*Separated* (as free acid) from sulphates and other substances insoluble in alcohol by its solubility in this menstruum (*d*); from tartaric acid, approximately, by the slight solubility of the potassic tartrate in dilute alcohol (*e*); from acids which form soluble lead salts by method given under Acetic acid at *g*.—*Determined* by acidimetry (*f*); by precipitation as barium citrate to be weighed as barium sulphate, or as barium citrate.

a. The citric acid of commerce is *crystallized* (from rather concentrated solutions) as $H_3C_6H_5O_7 . H_2O$, in large, transparent, colorless, and odorless prisms of the trimetric system. These crystals slowly effloresce in the air between 28° and 50° C., and lose all their water of crystallization at 100° C. A different form of crystals, containing one molecule of water to two molecules of acid, is obtained from boiling, concentrated solutions.—Citric acid melts when heated, and at 175° gives off pungent, characteristic vapors, containing acetone (see Acetic acid, 40, *c*), while Aconitic acid (9) is formed in the residue. (The odor is distinctly unlike that of heated Tartaric acid.)—Citric acid is *soluble* in less than its weight of water, in 1.5 parts of 90 per cent. alcohol, insoluble in absolute ether, but soluble to a slight extent in ether containing alcohol or water; also slightly soluble in chloroform containing alcohol.

b. The alkaline citrates are freely *soluble* in water; iron, zinc, and copper citrates, moderately soluble; other metallic citrates mostly insoluble, calcium citrate being somewhat soluble in cold water, but nearly insoluble in hot water. Ammonio-ferric citrate is readily soluble in water. Citric acid prevents the precipitation of iron and many other heavy metals by the alkalies, soluble double citrates being formed. The alkaline citrates are sparingly soluble in hot, less soluble in cold alcohol.—**Solution of lime,** added to solution of citric acid or citrates, causes no precipitate in the cold (distinction from Tartaric, Racemic, Oxalic acids); but on boiling a slight precipitate is formed (distinction from Malic acid). Solution of chloride of calcium does not precipitate solution of free citric acid even on boiling, nor citrates in the cold, but precipitates citrates (neutralized citric acid) when the mixture is boiled. The precipitate, $Ca_3(C_6H_5O_7)_2 . 2H_2O$, is insoluble in cold solution of potassa (which should be not very dilute and nearly free from carbonate), but soluble in solution of cupric chloride (two means of distinction from Tartaric acid); also soluble in cold solution of chloride of ammonium and readily soluble in acetic acid.—Solution of **acetate of lead** precipitates from solutions of neutral citrates, and from even very dilute alcoholic solution of citric acid, the white citrate of lead, $Pb_3(C_6H_5O_7)_2 . \frac{1}{2}H_2O$, somewhat soluble in free citric acid, soluble in nitric acid, in solutions of all the alkaline citrates and of chloride and nitrate of ammonium, soluble in ammonia (formation of basic citrate of lead then soluble with the citrate of ammonium produced). (Malate of lead is not soluble in malate of ammonium.)

c. **Nitrate of silver** precipitates from neutral solutions of citrates, white normal citrate of silver, not blackened by boiling (distinction from Tartrate).—Solution of **permanganate** of potassium is scarcely at all affected by free citric acid in the cold.

With free alkali, the solution turns green slowly in the cold, readily when boiled, without precipitation of brown binoxide of manganese till after a long time (distinction from Tartrate).

d. Citric acid is separated from "extractive matters" and from acids which form soluble barium salts by precipitation, as *barium citrate*, which is then carefully decomposed with sulphuric acid.—From citrates soluble in water, the acid may be obtained by decomposing with *sulphuric acid* (not added in excess), then removing the water by evaporation at a temperature below 100°, and extracting the citric acid from the residue by *alcohol.*

e. One part of citric acid dissolved in two parts of water, and treated with a solution of one part of acetate of **potassium** in two parts of water, will remain clear after addition of an equal volume of strong **alcohol** (absence of Tartaric, Racemic, and Oxalic acids). For a method by treatment of the crystals with alcoholic solution of potassa, see Tartaric acid (1), *c.*

Quantitative.—*f.* Uncombined citric acid, not mixed with other acids, may be determined *volumetrically* by adding a standard solution of soda or potassa to the neutral tint of litmus. Weighing 7.000 grammes ($\frac{1}{10}$ of $\frac{1}{3}$ of $C_6H_8O_7 \cdot H_2O$) the number of cubic centimeters of normal solution of alkali required equals the number per cent. of crystallized acid.*

g. The precipitation of alkaline citrates by **barium acetate** is made complete in solution of **alcohol** of sp. gr. 0.908—as follows :†

The citric acid is obtained as alkaline citrate; if free, by neutralization with soda; if combined with a non-alkaline base, by warm digestion with an excess of soda or potassa, filtering and washing—the filtrate being neutralized by acetic acid. In either case, the carefully neutralized and not very dilute solution is treated with a slight excess of exactly neutral solution of acetate of barium, and a volume of 95 per cent. alcohol, equal to twice

* Results a little too high.—J. CREUSE.

† J. CREUSE, *American Chemist*, I., 424 (1871).

that of the whole mixture, is added. The precipitate is washed on the filter with 63 per cent. alcohol, and dried at a moderate heat. The citrate of barium contains a variable quantity of water, and is transformed into sulphate of barium by transferring to a porcelain capsule, burning the filter, and heating with sulphuric acid several times, till the weight is constant. $3BaSO_4$: $2H_3C_6H_5O_7 . H_2O$: : 1 : 0.601.

Hager directs that barium or calcium citrate (washed with alcohol) be dried at 120° to 150° and weighed. $Ba_3(C_6H_5O_7)_2$: $2H_3C_6H_5O_7 . H_2O$: : 1 : 0.53232.

9. ACONITIC ACID. $H_3C_6H_3O_6$. A colorless solid, crystallizing with difficulty in warty masses, at 160° C. (320° F.) resolved into liquid itaconic acid. Soluble in water, alcohol, and ether; its solutions having a decided acid reaction. It has a purely acid taste.—The *aconitates* of the alkaline metals, magnesium, and zinc are freely soluble, the others insoluble or sparingly soluble, in water. Calcic aconitate is soluble in about 100 parts of cold water and in a much smaller quantity of boiling water. Manganous aconitate crystallizes in rose-colored octahedrons, sparingly soluble in water. Argentic aconitate is sparingly soluble in water, soluble in alcohol or ether, blackened by boiling with water.—Free aconitic acid is precipitated by mercurous nitrate, but not by most metallic salts until after neutralization.

Aconitic acid is *separated* from Monkshood (*Aconitum napellus*), Larkspur (*Delphinium consolida*), *Equisetum*, Black Hellebore, Yarrow (*Achillea millefolium*), and other plants, in which it exists as calcium salt, by evaporating the clear decoction to crystallize. The crystals of aconitate of calcium are dissolved and precipitated by acetate of lead, and the lead salt decomposed by hydrosulphuric acid. It is also separated from impurities by adding (to the dry mixture) five parts of absolute alcohol, then saturating the filtered solution with hydrochloric acid, and adding water, when aconitate of ethyl will rise as an oily layer, colorless and of aromatic odor. This ether may be decomposed by potassa.

Aconitic acid may be separated from Maleic acid by the more ready crystallization of the latter, and from Fumaric acid by being more soluble in water.

10. MALIC ACID. $H_2C_4H_4O_5$. *Identified* more especially by its deportment when heated (*a*); by the deportment of its lead salt when heated under water (*b*), and of its calcium salt in water and in alcohol (*d*).—*Separated* from tartaric, citric, oxalic, and other acids by alcoholic solubility of the neutral malate of ammonium (*c*) and by its reaction with calcium in water solutions (*d*); from tannic acid, also, by aqueous solubility of calcic malate, and from formic, acetic, benzoic acids by alcoholic insolubility of calcic malate (*d*).—*Determined* gravimetrically as lead malate (*e*).

Crystallizes in four-sided or six-sided prisms, deliquescent in air; colorless, odorless, and of sour taste; freely soluble in water and alcohol, soluble in ether. The malates are mostly soluble in water, but insoluble in alcohol. Nitric acid oxidizes malic acid, and alkaline solution of permanganate is decolorized by it, but chromic acid acts on it with difficulty. Malate of silver darkens but slightly on boiling (Tartrate blackens). Concentrated sulphuric acid darkens malic acid very slowly after warming. Hydriodic acid changes it to succinic acid with separation of iodine (the result being the same with Tartaric acid). Sodium amalgam changes malic to succinic acid.

a. Free malic acid, *heated* in a small retort over an oil-bath to 175° or 180° C., evolves vapors of maleic and fumaric acids, which crystallize in the retort and receiver. The fumaric acid forms slowly at 150° C., and mostly crystallizes in the retort, in broad, colorless, rhombic or hexagonal prisms, which vaporize without melting at about 200° C., and are soluble in 250 parts of water, easily soluble in alcohol or ether. If the temperature is suddenly raised to 200°, the maleic acid is the chief product.

Maleic acid crystallizes in oblique, rhomboidal prisms, which melt at 130° and vaporize at about 160°, and are readily soluble in water and in alcohol. The test for malic acid, by heating to 175° or 180°, may be made in a test-tube, with a sand-bath, the sublimate of fumaric and maleic acids condensing in the upper part of the tube. Malic acid melts below 100°, and does not lose weight at 120°; at the temperature of the test water-vapor is separated—maleic and fumaric acids both having the composition of malic anhydride ($C_4H_4O_4$).

b. Solution of **acetate of lead** precipitates malic acid, more perfectly after neutralizing with ammonia, as a white and frequently crystalline precipitate which upon a little boiling melts to a transparent, waxy semi-liquid (a characteristic reaction, obscured by presence of other salts). The precipitate is very sparingly soluble in cold water, somewhat soluble in hot water (distinction from Citrate and Tartrate); soluble in strong ammonia, but not readily dissolved in slight excess of ammonia (distinction from citrate and tartrate); slightly soluble in acetic acid.

c. If the precipitate of malate of lead is treated with excess of **ammonia,** dried on the water bath, triturated and moistened with alcoholic ammonia, and then treated with **absolute alcohol,** only the malate of ammonium dissolves (distinction from Tartaric, Citric, Oxalic, and many other organic acids, the normal ammonium salts of which are insoluble in absolute alcohol). Also, malic acid may be separated from tartaric, oxalic, and citric acids, in solution, by adding ammonia in slight excess, and then 8 or 9 volumes of alcohol, which leaves only the malate of ammonium in solution.

d. Solution of **chloride of calcium** does not precipitate malic acid or malates in the cold (distinction from Oxalic and Tartaric acids); only in neutral and very concentrated solutions is a precipitate formed on boiling (while calcic citrate is precipitated in neutral boiling solutions if not very dilute). The addition of **alcohol** after chloride of calcium produces a white bulky precipitate of calcic malate in even dilute neutral solutions (indicative in absence of sulphuric and other acids whose calcium salts are less soluble in alcohol than in water).—Acetic, Formic, and Benzoic acids are left in solution and malic acid precipitated by addition of one or two volumes of alcohol, with chloride of calcium. In separation from Tannic acid, both acids may be precipitated by chloride of calcium, with a slight excess of ammonia and alcohol; the malate is then washed out of the precipitate with water.

Quantitative.—*e.* The alcoholic solution of malate of ammonium—prepared as directed in *c*—may be precipitated with acetate of lead, washed with alcohol, dried and weighed as malate of lead. $PbC_4H_4O_5 : H_2C_4H_4O_5 :: 1 : 0.3953$.

11. MECONIC ACID. $H_3C_7HO_7$. *Identified* by its physical properties and precipitation by hydrochloric acid (*a*); its reactions with iron and other metals (*b*); and by its products when heated (*c*). It is *separated from opium* through formation of the calcium salt or lead salt (*d*).

a. Meconic acid crystallizes in white shining scales or small rhombic prisms, containing three molecules of crystallization water, fully given off at 100° C. It is soluble in 115 parts of water at ordinary temperatures, less soluble in water acidulated with hydrochloric acid, more soluble in hot water, freely soluble in alcohol, slightly soluble in ether. It has an acid and astringent taste and a marked acid reaction. Its salts, having two atoms of its hydrogen displaced by acid, are neutral to test-paper. Except those of the alkali metals, the dimetallic and trimetallic meconates are mostly insoluble in water. Meconates are nearly all insoluble in alcohol. They are but slightly or not at all decomposed by acetic acid.

Solutions of meconates are precipitated by **hydrochloric acid,** as explained above.

b. Solution of meconic acid is colored red by solution of **ferric chloride.** One ten-thousandth of a grain of the acid in one grain of water with a drop of the reagent acquires a distinct purplish-red color (WORMLEY). The color is not readily discharged by addition of dilute hydrochloric acid (distinction from Acetic acid), or by solution of mercuric chloride (distinction from sulphocyanic acid).—Solution of **acetate of lead** precipitates meconic acid or meconates as the yellowish-white meconate of lead, $Pb_3(C_7HO_7)_2$, insoluble in water or acetic acid.—Excess of **baryta** water precipitates a yellow trimetallic meconate.—Solution of **nitrate of silver** in excess precipitates free meconic acid on boiling, and precipitates meconates directly, as yellow trimetallic meconate; if free meconic acid is in excess, the precipitate is first the white dimetallic meconate; both meconates being soluble in ammonia and insoluble in acetic acid.—Solution of **chloride of calcium** precipitates from solutions of meconic acid, and even from neutral meconates, chiefly the white monometallic meconate, $CaH_4(C_7HO_7)_2 \cdot 2H_2O$, sparingly soluble in cold water; in the presence of free ammonia, the less soluble, yellow dimetallic salt, $CaH\,C_7HO_7 \cdot H_2O$, is formed. Both precipitates are soluble in about 20 parts of water acidulated with hydrochloric acid.

c. At 120° C. (248° F.) dry meconic acid is resolved into *comenic acid;* at above 200° C. the comenic acid is resolved into pyrocomenic acid and other products. The sublimate of comenic acid dissolves sparingly in hot water, not at all in absolute alcohol. It crystallizes in prisms, plates, or granules. Solution of comenic acid gives a red color with ferric chloride, green pyramidal crystals with cupric sulphate in concentrated solution, and a yellowish-white granular precipitate with acetate of lead.

d. The *separation of meconic acid from opium* is effected with least loss by precipitating the infusion with acetate of **lead** (leaving the alkaloids as acetates with some excess of lead in the filtrate). The precipitate is decomposed, in water, with hydrosulphuric acid gas, and the filtrate therefrom is concentrated (and acidulated with hydrochloric acid) to crystallize the meconic acid.

The crystals are purified by dissolving in hot water and crystallizing in the cold after acidulation with hydrochloric acid.

The **calcium** meconate, precipitated in concentrated solution by Gregory's process for preparation of morphia, as by the Br. Pharmacopœial preparation of morphiæ murias, is washed with cold water and pressed. One part of the precipitate is dissolved by digestion in 20 parts of nearly boiling water with 3 parts of commercial hydrochloric acid, and set aside to crystallize the acid meconate of calcium. The crystals are purified from color and freed from calcium by repeated solution in the same solvent, used just below 100° C., and each time in a slightly diminished quantity. The acid may be further decolorized by neutralizing with potassic carbonate, dissolving in the least sufficient quantity of hot water, draining the magma of salt when cold, dissolving again in hot water and adding hydrochloric acid to crystallize.

12. DIGITALIC ACID. Digitaleic acid. Digitoleic acid. —A solid of a green color, crystallizing in slender needles, sometimes stellate, having a bitter taste and pleasant aromatic odor. It is sparingly soluble in water, freely soluble in alcohol and in ether. It reddens litmus and decomposes carbonates. The alkaline digitalates are soluble, forming soapy solutions with water; the other metallic digitalates are insoluble in water.

Digitalic acid is obtained from the leaves of *the fox-glove* (digitalis purpurea) by exhausting with cold water, precipitating the solution with acetate of lead, decomposing the precipitate with solution of carbonate of sodium, and treating the somewhat concentrated filtered solution with hydrochloric acid, which precipitates crude digitalic acid. This is purified by crystallization from alcohol.

13. TANNIC ACIDS: Vegetable educts having an *astringent taste* and an acid effect on test-papers, mostly amorphous, not volatile or liquefiable, freely soluble in water and in alcohol, mostly but little soluble in dilute sulphuric acid; forming with **ferric salts** green or blue colors, and precipitating solutions of **gelatin** and albumen (distinctions from gallic acid). Attributed formulæ, $C_{27}H_{24}O_{18}$; $C_{27}H_{22}O_{17}$; $C_{14}H_{10}O_{9}$ (SCHIFF).

The tannic acids are further *characterized* by forming in solutions of all the caustic **alkalies** a brown color bleached by oxalic acid, and in solutions of **alkaloids** a white precipitate dissolved by acetic and stronger acids. With exceptions hereafter named, they precipitate solution of tartrate of **antimony** and potassium; they precipitate basic acetate of lead, and form insoluble compounds with many heavy metals. They all absorb oxygen, especially in presence of alkalies, and act as powerful *reducing* agents —quickly decolorizing solution of permanganate, and reducing the heated alkaline copper solution. Tannic acids are more permanent in alcoholic than in aqueous solutions.

If a very little starch-paste be tinged blue by a slight addition of hundredth-normal solution of **iodine** (1 part iodine dissolved with potassic iodide in 100,000 parts aqueous solution), on adding a liquid containing tannic acid the blue color of the iodized starch presently disappears—hydriodic acid and gallic acid being formed. On adding a crystal of potassic nitrite the blue is restored.* Also, if a drop of tannic acid solution is mixed with a few drops of iodine solution of the above strength, and afterward a drop of very dilute alkali be added, on evaporation to remove carbonic acid, a bright red color will appear.† By oxidation the tannic acids acquire a dark color, brown, black, green, or red. Gallotannic acid with alkalies in the air slowly forms tannoxylic acid, which precipitates acetate of lead solution dark-red. With lime-water it forms a white turbidity, becoming green and darker. Tannic acids form with **molybdate** of ammonium a red color removed by oxalic acid.

The *physiological* tannic acid (WAGNER, 1866) or quercitan-

* GRIESSMAYER: *Ann. Ch. Pharm.*, clx., 40–56.
† GRIESSMAYER: *Zeitschr. Anal. Chem.*, x., 43.

nic acid is found in the bark of the oak, pine, willow, and beech, in bablah (acacia fruit), in valonia (cups of the quercus ægilops), and in sumac. It is a glucoside, and it does not yield pyrogallic acid by dry distillation. The *pathological* tannic acid of Wagner, or gallotannic acid, is found in common or Turkish gall-nuts and in Chinese and Japanese gall-nuts. It is a glucoside (being transformed by contact of a ferment or by sulphuric acid into gallic acid and glucose), and in dry distillation it yields pyrogallic acid.

Ferric salts give *blue* to blue-black precipitates with gallotannic acid, quercitannic acid, and the tannic acids of poplar bark, birch bark, hazel-nut, uva ursi leaves, lithrum salicaria leaves, the bark of cornus florida and cornus mascula, and many other plants. Ferric gallotannate (*ink*) is bleached by oxalic acid. On digestion with nitric acid, a yellowish solution is formed, in which excess of ammonia precipitates ferric hydrate. Ferric salts give *green* precipitates with quinotannic acid, moritannic acid, caffetannic acid, catechutannic acid, catechuic acid, cephaelic acid, the tannic acids of the barks of pines and fir and willow, the rhubarb root, the root of potentilla tormentilla, and of numerous other plants. Cephaelic acid with ammonia is colored violet to black by ferric salts.

Gelatin does not precipitate Catechuic acid or Caffetannic acid.

Tannic acids are *removed from solution* by digestion with oxide of copper, oxide of zinc, or animal membrane; or by precipitation with solution of gelatin, sulphate of cinchonia, or acetate of copper.—They are *separated as insoluble lead salts*, according to the general method given under Acetic Acid.

Quantitative.—The total tannic acids in solution are determined—by the specific gravity (*a*); by absorption in oxide of copper (*b*); by a volumetric solution of sulphate of cinchonia (*c*); by a volumetric solution of tartrate of antimony and potassium (in presence of chloride of ammonium to prevent the precipitation of gallic acid) (*d*).

a. A water solution of gallotannic acid at 17.5° C. (63.5° F.) contains as follows (after HAGER):

P. C.	SPEC. GRAV.	P. C.	SPEC. GRAV.	P. C.	SPEC. GRAV.
20	1.0824	13	1.0530	6	1.0242
19.5	1.0803	12.5	1.0510	5.5	1.0222
19	1.0782	12	1.0489	5	1.0201
18.5	1.0761	11.5	1.0468	4.5	1.0181
18	1.0740	11	1.0447	4	1.0160
17.5	1.0719	10.5	1.0427	3.5	1.0140
17	1.0698	10	1.0406	3	1.0120
16.5	1.0677	9.5	1.0386	2.5	1.0100
16	1.0656	9	1.0365	2	1.0080
15.5	1.0635	8.5	1.0345	1.5	1.0060
15	1.0614	8	1.0324	1	1.0040
14.5	1.0593	7.5	1.0304	0.5	1.0020
14	1.0572	7	1.0283	0	1.0000
13.5	1.0551	6.5	1.0263		

When other substances besides tannic acid and water are present, the specific gravity of the solution is first taken; the solution is then deprived of tannic acid by digestion with animal membrane. Four to five parts of dried and rasped hide are added for one part supposed tannic acid. After digestion, the filtrate and washings are brought to the exact bulk of the original solution and to the standard temperature. The former specific gravity minus the latter, and plus one, equals the specific gravity indicating the per cent. of tannin. Gallic acid is not taken out by the membrane.—If pectous substances are present, they would also be precipitated by the animal membrane; hence they must be removed before taking the specific gravity in the first place. This is accomplished by making an alcoholic extract of the original material, then evaporating off the alcohol and substituting water (HAMMER).—Instead of animal membrane, oxide of copper may be used to remove the tannic acid (and gallic acid), according to *b*.

b. A weighed quantity of recently ignited oxide of copper—about 5 times that of the tannin—is added to the prepared solution; the mixture is gently warmed for an hour and set aside for a day with frequent agitation, then filtered and the copper tannate and oxide washed, dried on the water-bath and weighed. The increase of weight is the amount of tannic (and gallic) acid (HAGER).

c. 4.523 grams of good sulphate of cinchonia, with 0.5 gram sulphuric acid, and 0.1 gram acetate rosanilin or fuchsin, are dissolved in water to make one litre. Each c.c. of this solution precipitates 0.01 gram tannic acid. One gram of solid material is obtained in clear solution of about 50 c.c. measure. To this the standard solution of cinchona is added, the color being thrown down in the precipitate. When the tannic acid is all precipitated, the anilin color appears in solution. One gram having been taken, each c.c. of the volumetric solution indicates 1 per cent. of tannic acid. Gallic acid is not precipitated by cinchonia (R. WAGNER).

d. One equivalent of tartrate of antimony and potassium, after drying on the water-bath (**K SbO** $C_4H_4O_6$=325), is precipitated by one equivalent of tannic acid ($C_{27}H_{24}O_{18}$=636); or, 0.002555 anhydrous tartrate is precipitated by 0.005 of the tannin. Dissolving 2.555 grams of anhydrous tartrate of antimony and potassium in water to make one litre of solution, each c.c of the same corresponds to 0.005 of tannic acid. The prepared solution of tannic acid—which may contain pectous substances without interference with this method—is treated with chloride of ammonium, and the volumetric solution is added, with agitation, until turbidity is no longer produced. The precipitate separates well. Gallic acid is not thrown down when chloride of ammonium is present (GERLAND).

14. GALLIC ACID. $C_7H_6O_5$; crystallizing with H_2O. An inodorous solid, having an astringent and slightly acid taste, an acid effect on test-papers, and crystallizing in long, silky needles or triclinic prisms. It is soluble in 100 parts of cold or 3 parts of boiling water, freely soluble in alcohol, moderately soluble in ordinary ether, and but slightly soluble in absolute ether, insoluble in chloroform or petroleum naphtha. Its non-alkaline metallic salts are insoluble in water but soluble in alcohol, and slightly soluble in officinal ether; they are decomposed by acids and by alkalies.

Heated to 210°–215° C. (410°–419° F.), in absence of water, it is sublimed as pyrogallic acid and carbonic anhydride; at higher temperatures, other products are formed.

Gallic acid is *characterized* by its physical properties (as above given); by its reactions with iron salts (*a*), with alkalies (*b*), with tartrate of antimony and potassa and with alkaline arsenate in the air (*c*), and with molybdate of ammonium (*d*). It is *distinguished from the tannic acids* by negative results with gelatin, albumen, and starch (*e*); by not precipitating the alkaloids, and by its far weaker reducing power (*f*) (distinction from pyrogallic acid also).—Gallic acid is *determined*, if free from tannic acids, by absorption in recently ignited oxide of zinc, according to method *b* in determination of tannic acid. It is separated from tannic acids and determined by solution with carbonate of ammonium from the precipitate with acetate of copper (*g*).

a. Ferric salts in solution give a deep blue color with gallic acid. Ferrous salts give a blue-black precipitate (distinction from gallotannic acid).

b. Alkaline solutions of gallic acid turn yellow to brown and black in the air, from absorption of **oxygen** and formation of tannomelanic acid, greatly accelerated by boiling. The latter acid, on neutralizing with acetic acid, precipitates acetate of lead, black.

Solution of **lime** with gallic acid, forms a white turbidity, changing to blue and then to green.

c. Tartrate of **antimony** and potassium is precipitated white in very dilute solution.

A faintly alkaline solution of **arsenate** of potassium or sodium, with gallic acid, exposed to the air, soon develops an intense green color, commencing at the surface. Dilute acids change the green to purple-red and a careful neutralization with alkalies restores the green color, but it is destroyed by excess of alkali.*

* PROCTOR: *Jour. Chem. Soc.*, 1874, p. 509.

d. **Molybdate** of ammonium reacts as with tannic acid.

e. Gallic acid does not precipitate gelatin, albumen, or starch-paste, but it forms a precipitate with a mixture of gum-arabic and gelatin.

f. Gallic acid does not reduce alkaline copper solution, but *reduces* salts of gold and silver, and quickly decolorizes permanganate solution.

Quantitative.—*g.* The prepared solution is fully precipitated with a filtered solution of cupric acetate; the precipitate washed and then exhausted with cold solution of carbonate of ammonium. The last solution, containing all the gallate of copper with a very little tannate, is evaporated to dryness, the residue moistened with nitric acid, ignited, and weighed as oxide of copper. This weight multiplied by 0.9 gives the quantity of gallic acid (the full ratio being 0.9126, but allowance is made for solution of a little tannate by the carbonate of ammonium. The ratio between oxide of copper and tannic acid is 1.304). (Method of FLECK modified by SACKUR and WOLF.)

15. PYROGALLIC ACID. $C_6H_6O_3$. Pyrogalline. Pyrogallol.—*Characterized* by its physical properties (*a*); its peculiar avidity for oxygen (*b*); its reactions with alkalies, lime, iron, copper, etc. (*c*). It is *distinguished* from tannic acid by not precipitating gelatin or moderately dilute tartrate of antimony and potassium or cinchonia, and by its different reactions with both ferrous and ferric salts: from gallic acid by its greater solubility in cold water and its far greater reducing power (*b*). It may be determined gravimetrically as a lead precipitate (*d*), and volumetrically by permanganate.

a. Pyrogallic acid crystallizes in long prismatic plates or needles, of a white or yellowish-white color, and an acid and very bitter taste. At 115° C. (239° F.) it melts, and at about 210° C. (410° F.) it sublimes with partial decomposition and formation of metagallic acid. It is soluble in three parts cold water, freely soluble in alcohol and in ether, not soluble in absolute chloroform.

b. It is permanent in dry air free from ammonia, but in moist or ammoniacal air it gradually darkens, and in water solution it turns brown to black, sooner if boiled, still more rapidly in presence of alkalies, *absorption of oxygen* taking place to an extent proportional to the coloration, which is destroyed by oxalic acid. It quickly reduces the alkaline copper solution; also salts of the noble metals, and reduces acid solution of permanganate with evolution of carbonic anhydride.

c. With lime solution, a purple-red color at first appears, afterward the brown color formed by alkalies as mentioned in *b.* With ferrous salts a blue color is formed; with ferric salts a red solution, brown when heated. Acetate of copper gives a brown-green precipitate; acetate of lead a white, curdy precipitate; both soluble in acetic acid.

Quantitative.—*d.* The alcoholic solution of pyrogallic acid is precipitated with excess of alcoholic solution of acetate of lead; the precipitate washed quickly with alcohol, dried by water-bath and weighed. $Pb(C_6H_5O_3)_2 : 2C_6H_6O_3 :: 457 : 252 :: 1 : 0.55142.$

16. QUINOTANNIC ACID. Cinchotannic acid. Kinotannic acid.—See Tannic acids (13) for appearance, taste, solubilities, and reactions with alkalies and with iron salts. It precipitates tartrate of antimony and potassium only in concentrated solutions. In oxidation with alkalies it forms a red-brown color, due to cinchona-red, which dissolves in alkalies and in acetic acid, but not in water. Concentrated sulphuric acid changes quinotannic acid to cinchona-red and glucose. In dry distillation, phenic acid is formed, recognized by the odor. Quinotannic acid is removed from solution by acetate of lead, and from its lead precipitate by hydrosulphuric acid. For separation from Cinchona bark, see under Quinic Acid, *d.*

17. CATECHUTANNIC ACID. Has the properties of tannic acids in general, giving a grayish-green precipitate with ferric salts, and distinguished by not precipitating tartrate of antimony and potassium. It softens when heated, and by dry distillation yields an empyreumatic oil. The product of its atmospheric oxidation in water is red.

Catechutannic acid is *separated from Catechu* as follows: The aqueous infusion of catechu is heated with dilute sulphuric acid and filtered; the filtrate treated with concentrated sulphuric acid to precipitate the acid sought; the precipitate is washed on a filter with dilute sulphuric acid and pressed between paper. It may then be dissolved in water; the solution treated with carbonate of lead and filtered; the filtrate evaporated in vacuo. It may be farther purified by dissolving in alcoholic ether and evaporating off the solvent.

18. CATECHUIC ACID. Catechucic acid. Catechin. Tanningenic acid.—A white, tasteless powder, or in fine, silky needles, melting at 217° C. (423° F.), and in dry distillation yielding an empyreumatic oil. Very slightly soluble in cold water, soluble in three parts boiling water, moderately soluble in alcohol, sparingly soluble in ether. With alkalies and metallic salts, and as a reducing agent, it gives the reactions of the (iron-green) Tannic Acids, from which it is distinguished by not giving precipitates with tartrate of antimony and potassium or with alkaloids, or with gelatin (the last-named being a distinction from catechutannic acid). With strong sulphuric acid it forms a deep purple liquid.

Catechuic acid may be *separated from catechutannic acid and the other constituents of catechu* by its sparing solubility in cold and ready solubility in hot water. Bengal catechu is digested twenty-four hours in cold water, and the (slightly washed) residue is then exhausted with boiling water. When the solution cools, a yellow deposit of catechuic acid appears. This is washed in cold water. It may be decolorized by hot filtration through animal charcoal. It is dried on bibulous paper by aid of the air-pump.

19. MORINTANNIC ACID. $C_{13}H_{10}O_6$. Capable of crystallization; yellow, with great tinctorial power, and of an astringent, sweetish taste. Melts at 200° C., and at higher temperatures distils phenic acid. In reactions with alkalies, oxidizing agents, gelatin, tartrate of antimony and potassium, iron salts, etc., it behaves like other Tannic Acids (13). With ferric salts it gives a greenish precipitate; with acetate of lead a yellow precipitate; with sulphate of copper a yellowish-brown precipitate; with stannous chloride a yellowish-red precipitate.

It is *separated from Fustic* by spontaneous deposition from the concentrated decoction.

20. CAFFETANNIC ACID. Caffeotannic acid. Has in general the physical properties of the Tannic Acids, but is not incapable of crystallization. It melts when heated, and then gives the odor of roasted coffee, and in dry distillation yields oxyphenic acid as an oil which solidifies in the cold.

With fixed **alkalies** in solution it turns yellow to reddish-yellow, by oxidation; with ammonia, forms a green color, due to viridic acid, which, when neutralized, gives with acetate of lead a blue precipitate. Warmed with concentrated sulphuric acid, it dissolves with a blood-red color. Distilled with dilute **sulphuric acid** and **binoxide of manganese,** it evolves quinone—a pungent and irritating vapor, condensing to a golden-yellow to dingy-yellow, crystallizable substance, heavier than water, in which it is but slightly soluble when cold.

Caffetannic acid gives the green color with **ferric salts.** It reduces nitrate of silver, in the specular form, when heated. It is distinguished from the larger number of Tannic Acids by not producing precipitates with tartrate of antimony and potassium or with gelatin, but it precipitates cinchonia and quinia (distinction from Catechuic acid). It gives a yellow precipitate with **barium** salts.

By gradual addition of acetate of **lead,** in decoction of coffee, it is precipitated next after (the very little) citric acid. Decomposing the precipitate with hydrosulphuric acid, and evaporating the filtrate, it is obtained in impure, yellowish mass.

21. BOHEIC ACID. $C_7H_{10}O_6$. Boheatannic Acid. Amorphous, pale-yellow solid, caking by exposure to the air, melting at 100° C. to a waxy mass, very soluble in water and alcohol. Both aqueous and alcoholic solutions gradually decompose by evaporation in the air. It colors ferric salts brown. With **baryta,** in alcoholic solution, it forms a yellow precipitate, $BaC_7H_8O_6 \cdot H_2O$. With acetate of **lead,** in alcoholic solution, it forms a grayish-white precipitate, $PbC_7H_8O_6 \cdot H_2O$, which can be washed with alcohol and dried at 100° C.

It is *separated from the quercitannic acid, in black tea,* by precipitating the latter with acetate of lead in the boiling decoction, filtering; after twenty-four hours filtering again, and neutralizing the clear solution with ammonia, when the yellow basic salt is precipitated, $PbO.PbC_7H_8O_6$. The latter may be decomposed in alcohol by hydrosulphuric acid, and the filtrate concentrated in vacuum or over oil of vitriol.

22. QUINIC ACID. $C_7H_{12}O_6$. Kinic acid.—*Identified* by its physical properties and reactions (*a*); by its generation of quinone (*b*); by its reactions with a few metals (*c*).—*Separated* from cinchona bark, by crystallization from a solution freed from quinovic acid (*d*); from cinchona bark, coffee, or bilberry, by precipitating its calcium salt from a sufficiently purified solution by adding alcohol (*e*); from substances forming insoluble compounds with neutral acetate of lead by the solubility of its normal lead salt.—*Determined* gravimetrically as calcium salt (*c*).

a. Colorless, monoclinic prisms or prismatic tablets, melting at 161° C. (322° F.), at higher temperatures evolving combustible gas, phenic acid, hydroquinone, etc. It is freely soluble in water, slightly soluble in alcohol, nearly insoluble in ether. Its solutions have a sour taste and redden litmus. It is deliquescent.

b. Distilled with moderately dilute sulphuric acid and binoxide of manganese, it yields an abundant yellow crystalline sublimate of *quinone*, recognized in very small quantities by its irritating odor, exciting tears. Farther, aqueous solution of Quinone is colored brown by ammonia, and yellow-green by chlorine water; it stains the skin brown.

c. Quinic acid decomposes carbonates. Its *metallic salts* are soluble in water, except the basic quinate of lead, but are insoluble in alcohol. It prevents the precipitation of many metallic oxides by alkalies. Quinate of silver is white, and bears the heat of the water-bath. The *quinate of calcium* crystallizes well from water solution as $\mathbf{Ca(C_7H_{11}O_6)_2 \cdot 5H_2O}$, which loses all its water of crystallization at 120° C. (248° F.) Or, it may be precipitated from solution of alkaline quinates by adding chloride of calcium, ammonia, and alcohol. The basic quinate of lead is precipitated by adding, to solution of alkaline quinate, basic acetate of lead, or normal acetate of lead with ammonia. It is somewhat soluble in solution of basic acetate of lead. It is variable and instable in composition.

d. The aqueous solution obtained by macerating *cinchona bark* two or three days (and from which the alkaloids may have been removed by acidulating with hydrochloric acid and then adding an excess of soda and, after a few hours, filtering) is treated with solution of acetate of lead to precipitate the quinovic and quinotannic acids, and filtered. The filtrate is evaporated to a syrupy consistence, to crystallize the quinic acid.

If it be desired to separate the Quinovic acid, the solution of acetate of lead (as above) is not added to complete precipitation, and the precipitated quinovate of lead is decomposed, in water, by adding very dilute sulphuric acid, drop by drop, with great care, to avoid excess. The precipitate being removed, the filtrate is concentrated for crystallization of the quinovic acid.

If the Quinotannic acid is to be obtained, the precipitation by acetate of lead is left incomplete, as directed next above, and the filtrate concentrated as previously directed for quinic acid.

With the crystals of quinic acid there will now finally deposit amorphous or oily quinotannic acid. This may be separated by washing with ether; on evaporation of the ether the quinotannic acid is obtained. [Thesis of R. M. COTTON, Univ. of Mich., 1874.]

e. After precipitating the alkaloids from decoction of *cinchona bark* with lime, according to the United States Pharmacopœial preparation of quiniæ sulphas, the filtrate is concentrated to a small bulk, filtered if necessary, and then alcohol is added to precipitate quinate of calcium. Or, the filtrate is concentrated to a soft solid, washed repeatedly with alcohol, and dissolved in enough water to allow the quinate of calcium to crystallize.

Fresh *bilberry plant* (vaccinium myrtillus), collected in May, is boiled with water and lime; the solution is evaporated, and alcohol added to precipitate the quinate of calcium, which requires purification by recrystallization from water.

Thoroughly dried or moderately roasted *coffee beans,* coarsely powdered, are exhausted by boiling with water; the decoction, mixed with milk of lime, is concentrated, filtered, evaporated on a water-bath to a syrup, and precipitated with alcohol as above.

The *quinate of calcium* obtained from any of the above sources may be purified from tannic acids and some coloring matters by adding solution of neutral acetate of lead to the aqueous solution of quinate of calcium, filtering out the lead precipitate, and removing the excess of lead from the filtrate by hydrosulphuric acid, when the last filtrate may be concentrated to crystallize. Quinic acid may be obtained from quinate of calcium by precipitating the aqueous solution of the latter by basic acetate of lead, and removing the lead from the precipitate by hydrosulphuric acid.

23. QUINOVIC ACID. $C_{30}H_{48}O_8$. Kinovic Acid. *Quinovin* or Kinovin. Quinova bitter or Kinova bitter.—An amorphous solid, having a very bitter taste, nearly insoluble in water, very soluble in alcohol, slightly soluble in ether, soluble in chloroform. (According to DE VRIJ, chloroform dissolves from "quinova bitter" a portion which he designates as "quinovin," leaving "quinovic acid" insoluble in that menstruum and little soluble in alcohol.) Dry hydrochloric acid gas, acting on a strong alcoholic solution of quinovic acid, transforms the latter into an acid and a sugar. The new acid has very nearly the same solubilities as the original acid, but a different composition ($C_{24}H_{38}O_4$), and forms definite salts with metals.

Quinovic acid forms a soluble calcium salt, and hence it is dissolved *from cinchona bark* by boiling with milk of lime. From the solution, sufficiently concentrated, hydrochloric acid separates the quinovic acid, insoluble in water. It may be purified by crystallization from alcohol, or by repeated precipitation from alcohol by water. For the separation of quinovic, quinic, and quinotannic acids, each from the same portion of bark, see Quinic Acid, *d*. In the manufacture of cinchona alkaloids, the acidulation of the water by which the decoction is made interferes with the solution of quinovic acid, which may be at least partly left in the residue.

24. COLUMBIC ACID. $C_{42}H_{40}O_{13}$. Colombic acid.—An amorphous solid, precipitated in white flakes, left as a yellowish, varnish-like residue on evaporation of its solutions. It is soluble in alcohol, nearly insoluble in water or ether, its solution being markedly acid. It is precipitated by neutral acetate of lead, as $(PbO)_3(C_{42}H_{44}O_{12})_2$ when dried at 130° C. Acetate of copper does not precipitate it.

In *columbo root*, columbic acid probably exists in combination with berberina and perhaps also with inorganic bases. It can be separated by exhausting alcoholic extract of columbo with water or lime-water, and precipitating with hydrochloric acid.

25. GENTIANIC ACID. $C_{14}H_{10}O_5$. Gentisic acid. Gentianin. Gentisin.—Light-yellow, tasteless, solid, crystallizing in slender needles, not decomposed at 200° C., but carbonizing with partial sublimation at 300° to 400° C. It is soluble in 36 parts water at ordinary temperature, readily soluble in alcohol, and moderately so in ether. Its solutions are neutral to litmus. It dissolves in aqueous alkalies with a golden-yellow color. Strong sulphuric acid dissolves it yellow. Nitric acid, of specific gravity 1.42, and colorless, dissolves it green; on adding water, a green powder, dinitrogentianic acid, is precipitated. This, on addition of alkalies, assumes a fine cherry-color. Chlorine forms a yellow precipitate in alcoholic solution of gentianic acid. The barium salt, **Ba $C_{14}H_8O_5 . H_2O$**, is an orange-colored precipitate. The lead salt is insoluble.

Gentianic acid is separated from *gentian root* as follows: The powdered root is exhausted of gentian-bitter by cold water; then pressed, dried, and exhausted with strong alcohol, and the alcoholic solution evaporated nearly to dryness. The residue is washed with a little ether to remove fat, and repeatedly crystallized from alcohol to separate from resin.

26. CARMINIC ACID. $C_9H_{20}O_6$. Carmine.—A purple amorphous solid, fusible but not decomposed at 136° C.; soluble in all proportions in water and alcohol, and in sulphuric and hydrochloric acids without alteration, the solutions having a bright purple-red color. Ether does not dissolve it.—In alcoholic solution it precipitates alcoholic potassa red changing to dark violet, and forms red precipitates with acetates of lead, zinc, copper, and silver. It is turned blue by sulphate of aluminum, and yellow by stannous chloride.—Carminic acid is a glucoside, boiling dilute mineral acids transforming it into *carmine-red* and sugar. Carmine-red in mass is purple-red with a green reflection, soluble in water and in alcohol with red color, not soluble in ether.

Carminic acid is *separated* from Cochineal by exhaustion with boiling water; the solution precipitated by adding slightly acidulated subacetate of lead short of excess, the precipitate washed with water till the washings give no precipitate with mercuric chloride, then decomposed by hydrosulphuric acid and filtered. The filtrate is evaporated and dried on the water-bath, and the residue extracted with alcohol.

27. CHRYSOPHANIC ACID. Chrysophane. Rheic Acid.—A pale yellow or orange-yellow solid, crystallizing in six-sided tables or moss-like aggregates of scales, subliming with partial decomposition when heated.—Sparingly *soluble* in cold water, soluble in 1,125 parts of 86 per cent. alcohol at 30° C. (86° F.), or 224 parts of the same alcohol boiling, soluble in ether, benzole, and turpentine oil, the solutions having a yellow color.—It dissolves in aqueous *alkalies* with a very deep purple color, recognized in very dilute solution; the potassa solution upon evaporation deposits violet to blue flocks, which dissolve in water to a red solution.—It does not form stable *salts*. In alcoholic solution with alcoholic subacetate of lead it forms a reddish-white precipitate, becoming rose-red when boiled with water. In ammoniacal solution it is precipitated lilac by neutral acetate of lead, and rose-color by alum.—Strong sulphuric acid dissolves it unchanged; strong nitric acid converts it into a red substance, containing chrysammic acid (produced from Aloes by nitric acid).

Chrysophanic acid is *separated* from Rhubarb by exhausting the powdered root with alcoholic ammonia, precipitating with subacetate of lead and decomposing the lead compound by hydrosulphuric acid. From the Wall Lichen (*Parmelia parietina*), the alkaline solution obtained as above is precipitated by acetic acid, the precipitate washed with water, redissolved in alkali and reprecipitated by (hydrochloric) acid. From the Rumex, an ethereal extract is obtained, and repeatedly dissolved in alcohol and precipitated by water. A method of purification is to dissolve in boiling absolute alcohol and crystallize.

28. GAMBOGIC ACID. A resinous solid, hyacinth-red in mass, yellow in powder. Insoluble in water, soluble in alcohol, ether, chloroform, bisulphide of carbon—its solutions showing the yellow color when very dilute, and having a strong acid reaction. It dissolves in the aqueous alkalies, with red color, and in solutions of fixed alkaline carbonates with expulsion of the carbonic anhydride. From alkaline solutions it is precipitated yellow by acids.—The solution of gambogiate of ammonia forms with barium salts a red precipitate; with zinc salts, yellow; lead salts, reddish-yellow; silver salts, brownish-yellow; and copper salts, brown precipitates.—It is bleached and decomposed by chlorine, and decomposed with formation of nitrophenic acid by nitric acid. It is dissolved with red color by cold concentrated sulphuric acid; addition of water precipitating it unchanged.

29. SANTALIC ACID. Santalin.—A fine red, tasteless, and odorless crystallizable solid, melting at 104° C. Insoluble in water, very soluble in alcohol, soluble in ether—the solutions having a blood-red color and acid reaction. Soluble in aqueous potassa, or ammonia, forming violet solutions, which precipitate alkaline earths.—The alcoholic solution precipitates lead salts, but not salts of barium, silver, or copper. The lead and barium salts are violet.

Santalic acid is *separated* from Sandal-wood (red saunders) by obtaining, first, an ethereal extract, then from this an alcoholic extract, which is washed with water, dissolved again in alcohol, and precipitated therefrom by alcoholic solution of acetate of lead. The lead compound is washed by alcohol, then decomposed in alcohol with dilute sulphuric acid.

SOLID VOLATILE ACIDS.

30. BENZOIC ACID. $HC_7H_5O_2$. *Identified* by its physical properties, especially in sublimation (*a*); by its oxidation to nitrobenzole (*b*), and its deoxidation to bitter almond oil (*c*); by its reactions with metallic salts (*d*).—*Distinguished* from

Cinnamic acid by the action of permanganate upon the latter (see 31, *b*); from Hippuric acid by distillation with potassa; from Salicylic acid by the color of its ferric salt (*d*).—*Separated* from non-volatile and highly volatile substances by sublimation (*a*); from Succinic and many other acids by the alcohol solubility of its barium salt (*d*); from Succinic and Hippuric acids by its solubility and extraction from water solutions by chloroform or ether (*c*).—Gravimetrically *determined* as lead salt (*e*).

a. A white solid, crystallizing in lustrous scales or friable needles; odorless when pure, but frequently found having odor of benzoin, and rarely a urinous odor, of an acid and warm taste, and a strongly acid reaction. It is soluble in 200 parts of water at 15° C. (59° F.), in 20 parts of boiling water, in 3 parts of cold alcohol, in 25 parts of ether, in 7 parts of chloroform, and readily soluble in bisulphide of carbon, benzole, petroleum naphtha, and in fixed and volatile oils. Most of the benzoates are soluble in water, and many of them are soluble in alcohol.

Hydrochloric acid precipitates benzoic acid from solutions of benzoates, excess of the reagent not affecting the water solubility of benzoic acid as already given. Sulphuric acid dissolves benzoic acid. Benzoic acid decomposes carbonates.

Benzoic acid melts at 121° C. (250° F.), and *sublimes* at 240° to 250° C. (464° to 482° F.) The vapors cause a sense of irritation in the throat and coughing. When slowly condensed, the sublimate is crystalline in minute needles. Benzoates heated with phosphoric acid evolve benzoic acid.—When mixed with 3 parts slaked lime and heated gradually in a retort, benzole (119) is distilled.

b. If benzoic acid is boiled with concentrated **nitric acid,** the mixture evaporated to a small bulk, and then strongly heated in a test-tube, *nitrobenzole* (120) is evolved, and will be recognized by its odor of bitter almond oil.

c. When benzoic acid, dissolved or suspended in water, is warmed with a slip of metallic **magnesium,** and very slightly acidulated with sulphuric acid, so that hydrogen is evolved, *bitter almond oil* (benzoyl hydride, C_7H_5OH) is produced, and recognized by its odor.

d. Basic **ferric chloride** solution precipitates benzoates almost completely, as a flesh-colored basic benzoate (ferric Salicylate is blue violet).—Acetate of **lead** and nitrate of **silver** give precipitates in solutions not too dilute.—Ammoniacal chloride of **barium** with alcohol gives no precipitate (distinction and separation from Succinic and many other acids). **Magnesium** benzoate is also soluble in alcohol (Succinate insoluble in alcohol).

Quantitative.—*e.* Benzoate of lead, precipitated from neutral benzoate by acetate of **lead,** washed with cold water or alcohol acidulated with one-half per cent. of acetic acid, and dried at 100° C., may be weighed for determination of benzoic acid: $Pb(C_7H_5O_2)_2 : 2HC_7H_5O_2 : : 1 : 0.54343$.

31. CINNAMIC ACID. $HC_9H_7O_2$. *Characterized* by its physical properties and reactions in the dry way (*a*); its reactions with oxidizing agents (*b*); its reactions with metallic salts (*c*).—*Distinguished* from benzoic acid by action with oxidizing agents (*b*), by the color of its ferric salt and by its precipitate with manganous salts (*c*).—*Separated* from non-volatile substances by sublimation (*a*); from substances soluble in water and in dilute acid by precipitation of cinnamates by acids (*a*); from substances insoluble in ether by the action of that solvent; from benzoic acid by manganous precipitation (*c*).

a. A colorless solid, crystallizing (from vapor or solution) in monoclinic prisms or laminæ, melting at 129° C. (264° F.), vaporizing at about 300° C. (572° F.) It is very sparingly soluble in cold, moderately soluble in boiling water, freely soluble in alcohol and in ether. The cinnamates of the alkali metals are soluble in water, those of the alkaline earthy metals sparingly soluble, the other cinnamates mostly insoluble, the silver salt nearly insoluble.

It is precipitated by water from its alcoholic solutions, and by hydrochloric acid from water solutions of its salts of alkali metals.

When slowly distilled, cinnamic acid evolves cinnamene, having a persistent aromatic odor resembling that of benzole and naphthalene together. Cinnamates subjected to dry distillation emit the odor of bitter almond oil.

b. A saturated hot-water solution, acidulated with sulphuric acid, is treated with a few cubic centimetres of a one per cent. solution of **permanganate** of potassium and warmed a few minutes. If cinnamic acid is present, the odor of bitter almond oil becomes apparent.—**Nitric acid** with gentle heat, **peroxide of lead** in boiling solution, **chromate and sulphuric acid** with heat, evolve bitter almond oil (hydride of benzoyl) from cinnamic acid—in most cases with simultaneous production of benzoic acid.—*Cinnamates* with strong **nitric acid** give off odor of cinnamon oil and bitter almond oil.

c. **Ferric** salts with cinnamates give a yellow precipitate; **manganous** salts with excess of cinnamates give a white precipitate (none with benzoates); **copper** salts, a greenish-blue precipitate; acetate of **lead,** a precipitate not soluble in water, $\mathbf{Pb(C_9H_7O_2)_2}$, from which alcohol washes out a part of the cinnamic acid; nitrate of **silver,** a stable white precipitate, $\mathbf{AgC_9H_7O_2}$, insoluble in boiling water; baric and calcic salts, precipitates, easily soluble in hot water.

32. SUCCINIC ACID. $\mathbf{H_2C_4H_4O_4}$. *Characterized* and identified by its physical properties (*a*); its resistance to oxidation (*b*); its reactions with iron, manganese, lead, barium, calcium, etc. (*c*).—*Distinguished* from cinnamic acid by the color of its iron salt and by not precipitating manganous salts (31, *f*).—*Separated* from non-volatile materials by sublimation (*a*); from benzoic acid by insolubility of its barium salt in alcohol (30, *d*), and by its insolubility in chloroform or ether; from cinnamic acid by the solubility of manganous succinate (30, *c*).—*Determined* by extraction with ammonia from the ferric succinate (*d*).

a. Crystallizes in the monoclinic system, generally rhombic or hexagonal plates. At 130° C. (266° F.) it begins to emit suffocating vapors, at 130° C. (356° F.) it melts, and at 235° C. (455° F.) it sublimes as succinic anhydride (**$C_4H_4O_3$**), which melts at 120° C. (248° F.) The succinic acid of commerce has usually more or less of yellow to brown color, and of the empyreumatic and slightly aromatic odor of oil of amber; when pure it is white, and at ordinary temperatures odorless. Succinic acid is soluble in about 13 parts of water at ordinary temperatures, in 2½ parts of hot water, in 30 parts of cold or 20 parts of boiling alcohol, sparingly soluble in ether, not soluble in chloroform or benzole.—Succinic *anhydride* is more soluble in alcohol, but less soluble in water than the acid.—The *succinates* of the alkali-metals and magnesium are soluble in water; of the alkaline earth-metals, and of most other metals in diatomic salts, sparingly soluble; ferric succinate, insoluble.

b. Nitric acid, chromic acid, and chlorine are without action upon succinic acid. Cold permanganate solution does not affect free succinic acid, but with free alkali oxalic acid is formed with deposition of binoxide of manganese.

c. **Ferric** chloride, better if slightly basic, precipitates from solutions of succinates a brownish-red bulky precipitate of basic ferric succinate.—**Manganous** salts do not precipitate succinates. —Acetate of **lead** and nitrate of **silver**, each, give white precipitates of normal succinates slightly soluble in water.—Ammoniacal chloride of **barium** with alcohol produces a white precipitate even in dilute solutions.

Quantitative.—*d.* The ferric succinate is precipitated from dilute solution of succinate by addition of ferric chloride, then acetate of sodium in excess, and then sufficient ammonia to nearly or quite neutralize the mixture. After boiling one-fourth of an hour, the precipitate is filtered out and washed, then boiled with excess of a five per cent. solution of ammonia and filtered and washed with ammoniacal water. The filtrate is evaporated on the water-bath until it ceases to lose weight, and weighed as **$NH_4HC_4H_4O_4$**. Or, for greater exactness, this salt while in solution is treated with a weighed quantity of recently calcined magnesia, and the mixture evaporated and dried at 150° C. (302° F.) The increase of weight represents the succinic anhydride.

33. SALICYLIC ACID. $C_7H_6O_3$. (In most salts of this acid one atom of hydrogen, in a few salts two atoms, are replaced by metals.)—Crystalline, in monoclinic four-sided prisms or slender needles. Melts at 125° to 150° C. (257° to 302° F.) and sublimes at about 200° C. (392° F.) Its vapor causes irritation in the throat: it has a sweetish-sour taste. It has a decided acid reaction upon test-papers.—It is slightly soluble in cold, moderately soluble in hot water, freely soluble in alcohol and in ether. —The salicylates of the alkali metals are insoluble in water; those of the alkaline earth metals sparingly soluble (that of calcium least); many of those of other metals not soluble. The dimetallic salts are less soluble than the monometallic.—With ferric salts, salicylic acid forms a deep violet color.

Distilled or heated with **methylic alcohol** and concentrated **sulphuric acid**, salicylate of methyl is evolved, having the odor of wintergreen oil.

34. VERATRIC ACID. $C_9H_{10}O_4$. Crystallizes in slender speculæ or four-sided prisms, which effloresce at 100° C. and melt at a higher temperature, then subliming without decomposition. It is sparingly soluble in cold, freely in hot water, soluble in alcohol, insoluble in ether—the solutions having a slight acid reaction. The alkaline veratrates are soluble in water and crystallizable; the lead and silver salts insoluble. It dissolves in concentrated nitric acid, and when this solution is diluted it deposits nitroveratric acid, soluble in alcohol, from which it crystallizes in yellow laminæ.

Veratric acid is *separated from sebadilla seeds* (veratrum sabadilla) as follows: They are exhausted with alcohol acidulated with sulphuric acid, the solution is precipitated with milk of lime and filtered, the filtrate—containing veratrate of calcium—is concentrated, treated with hydrochloric acid, and left in a cold place to crystallize. The crystals may be purified by dissolving in alcohol, and filtering through animal charcoal.

35. PHENIC ACID. HC_6H_5O. Purified Carbolic acid. Phenol. Phenylic alcohol. Coal-tar creosote.—*Characterized* and *identified* by its physical properties (*a*); by its reactions with nitric acid (*b*), with ferric salts (*c*), with bromine (*d*) and chlorine (*e*), as a reducing agent (*f*), and with sulphuric acid (*g*).—*Distinguished* from Creosote by reacting with ferric salt in more dilute solution (*c*), by gelatinizing collodion, by greater solubility in ordinary glycerin, in bisulphide of carbon, and in ammonia water, and by crystallizing when pure (*a*).—*Separated* from Cresylic acid and other constituents of crude carbolic acid or from Fats by its greater solubility in water (*h*); from solution (in a greater quantity of) water by saturation with common salt (*i*); from admixture with (a smaller quantity of) water or with other substances by treatment with chloroform or bisulphide of carbon (*j*); from Creosote, in part, by solution in water (*h*); from soaps by successive treatment with acid, water, and chloroform (*k*); from fixed and volatile oils by hot water.

a. Phenic acid is a colorless-white solid, crystallizing in long needles of the trimetric system, melting at 34° to 41° C. (93° to 106° F.), and distilling at 182° to 186° C. (359° to 367° F.) It has a strong and persistent odor, resembling creosote but somewhat aromatic, a biting taste, and (when concentrated) a bleaching and shrivelling effect on the skin. It does not redden litmus. —It is soluble in 20 parts of water at ordinary temperatures, and dissolves two or three per cent. of water, being thereby liquefied —hence is deliquescent in the air. It is soluble in all proportions of alcohol, ether, chloroform, bisulphide of carbon, and glycerin (absolute or ordinary); in 20 parts of benzole; readily soluble in fixed oils and many volatile oils, and in aqueous solutions of potassa and soda.—The last-named mixtures or compounds, sometimes termed phenates, are not of definite proportions, but are crystallizable, and are soluble in alcohol and ether. Phenic acid does not decompose carbonates, but mixes with aqueous alkaline carbonates.—It coagulates **albumen** and **gelatin** and **collodion** (ether-solution of gun-cotton).

b. To a few drops (or a small fragment) of the material to be tested add a drop or two of concentrated **nitric acid.** Then add a slight excess of potassa, and if color has appeared dilute with water. The yellow color of nitrophenic acid (36, *a*) is apparent in 10,000 parts of water; of the potassic nitrophenate in 50,000 parts of water; the column having the depth of half an inch.* The nitrophenic acid may be extracted from water by benzole or ether.

c. Very dilute solution of **ferric** chloride gives a blue color with aqueous solution of phenic acid—the color being permanent (distinction from that of Morphia), but destroyed by boiling (distinction from that of Tannic acid). Oxalic acid destroys the color, and many organic substances prevent its formation; it is not extracted by benzole or chloroform.

In this test, the result is distinguished from a similar one with Creosote by the following precautions (FLUCKIGER): 1st, take 1 part of solution of ferric chloride of specific gravity 1.34, and 9 parts of the liquid to be tested (with pure carbolic acid the mixture has a yellowish hue; with pure creosote, no color). 2d, add 5 parts of 85 per cent. alcohol (with pure carbolic acid, a clear brown liquid; with pure creosote, a green solution). 3d, add 60 parts of water; with pure creosote, the result is a dingy brownish color; if phenic acid is present, a fine blue color appears.

d. **Bromine** water gives a yellowish-white precipitate in even very dilute solutions of phenic acid (the same with Creosote).

e. **Chlorine** gas (from chlorate of potassium and hydrochloric acid) forms a deep yellow color—chloride of phenyl.

* PRESCOTT: Proceedings Am. Phar. Asso., **xix.**, 550, and Chem. News, xxvi., 269.

f. Alkaline cupric solution is not *reduced* by (pure) phenic acid (is reduced by crude "carbolic acid"). Mercury and silver salts are only slowly reduced by boiling with phenic acid (are reduced by impure).

Permanganate solution is reduced by pure phenic acid, in solutions acid or alkaline, with separation of binoxide of manganese.

g. With sulphuric acid—equal parts of the concentrated acids at 290° C. for a quarter of an hour furnishing the best result—Sulphophenic acid is formed (37).

Quantitative.—*h.* Cresylic acid and other admixtures (as fats) nearly or quite insoluble in water may be approximately separated and determined by solution with 20 parts of **water.** In a cylindrical graduate of $\frac{1}{4}$ litre (or larger) capacity, place 10 c.c. of the carbolic acid or mixture under examination, add 200 c.c. of water, agitate, and set aside. Read from the bottom the number of c.c. of impurities.

i. Phenic acid may be approximately *separated from water* solution by adding **chloride of sodium** as long as the latter dissolves. If the operation be performed in a cylindrical graduate, as above, the layer of phenic acid is read from the top.

j. Phenic acid may be approximately deprived of water and the amount of the latter ascertained by mixture with **chloroform** or **bisulphide of carbon.** In a graduate of a little more than 20 c.c. capacity (a test-glass or test-tube may be graduated for the purpose), place 10 c.c. of the phenic acid under examination and add 10 c.c. of the chloroform or bisulphide of carbon, agitate, stopper, and set aside a few hours. Read off from the top the amount of water separated.—Phenic acid may be separated from various mixtures in the same manner; for this purpose the mixture should be made neutral to test-paper, if not so already. The chloroform or bisulphide of carbon may be removed by evaporation in a warm place.

k. In separation of phenic acid *from soaps*, the latter is decomposed by digestion with dilute sulphuric acid and hot water; when cold, the fat acid is separated, by use of a wet filter if necessary, and washed with water; and the water solution and washings exhausted with chloroform. The chloroform may be distilled from the phenic acid, and if necessary the distillation repeated.

36. NITROPHENIC ACID. $HC_6H_2(NO_2)_3O$. (Trinitrophenic acid.) Trinitrophenol. Carbazotic acid. Picric acid.—*Identified* by its physical properties, especially its intense coloring effects (*a*); its precipitation of alkaloids (*b*); its reactions with special reagents (*c*).—*Separated* from water solutions by extraction with chloroform, etc. (*a*); by crystallization as a potassium salt (*d*).—*Determined* as salt of cinchonia (*e*).

a. In bright yellow crystalline scales or in octahedrons of the trimetric system. It melts when slowly heated and afterward sublimes; when quickly heated it explodes. It has a very bitter and somewhat acrid and sour taste, and when heated a suffocating odor and effect. It reddens litmus.

It is *soluble* in 100 parts of water at 15° C. (59° F.) and in 25 parts at 80° C. (176° F.), less soluble in water acidulated with mineral acids, and freely soluble in alcohol, ether, chloroform, benzole, petroleum naphtha, and amylic alcohol. These solvents, which are not miscible with water, remove nitrophenic acid from water by aid of acidulation with sulphuric acid. The solutions have a yellow color, perceptible when very dilute; except solutions in benzole, petroleum naphtha, and dilute sulphuric acid, which are colorless.

The colorless as well as colored solutions *stain* white paper, and more permanently stain the skin and fabrics of nitrogenous composition.

The normal metallic *picrates* are all soluble in water, that of **potassium** being one of the least soluble, and requiring 260 parts of cold or 14 parts of boiling water for solution. This salt is insoluble in alcohol.—Many of the picrates *explode* more violently than the free acid, and oxidizable agents in intimate contact facilitate explosion, which may occur by trituration or pressure.

b. Solution of salts of most of the **alkaloids** precipitate nitrophenic acid or its soluble salts—the cinchona alkaloids, the opium alkaloids, except morphia and pseudomorphia, the strychnos alkaloids, veratria, berberina, colchicia, and delphinia, being fully precipitated from solution even when dilute and well acidulated with sulphuric acid. Morphia is precipitated from moderately concentrated solutions having little or no free acid. The precipitates are yellow, and are dissolved by hydrochloric acid. Compare 135, *e*.

c. With ammoniacal **cupric** sulphate solution, nitrophenic acid forms a green precipitate.—Potassic cyanide, or potassic sulphide, or grape sugar, with nitrophenic acid and excess of potassa, in hot solution, gives a blood-red solution (yellow when greatly diluted) from formation of *isopurpurate* of potassium (the crystals of which are green by reflected light).—If ferrous sulphate is boiled in solution with nitrophenic acid, treated with excess of ammonia and filtered, the filtrate concentrated and acidulated with acetic acid, bright-red crystals of *picramic acid* are formed. Stannous chloride and several other reducing agents may be substituted for the ferrous salt. Picramic acid is nearly insoluble in water, but soluble in alcohol or ether.

d. The graded solubility of potassic nitrophenate in hot and cold water and in alcohol (*a*) enables this salt to be almost perfectly removed from solution, in beautiful crystals, by gradual cooling of the hot water solution, with gradual addition of alcohol after crystallization has ceased in the cold water.

Quantitative.—*e*. Nitrophenic acid or a soluble salt of this acid is precipitated by a solution of sulphate of **cinchonia** acidulated with sulphuric acid, the precipitate is washed with water, dried at a very gentle warmth, then heated (and melted) on the water-bath and weighed. $C_{20}H_{24}N_2\ (C_6H_2[NO_2]_3O)_2 : 2HC_6H_2(NO_2)_3O :: 1 : 0.6123$.

37. SULPHOPHENIC ACID. $HC_6H_5SO_4$. Phenyl sulphuric acid. Sulphophenylic acid. Sulphocarbolic acid.—Only preserved in its salts, which are stable and crystallizable compounds, decomposed by **nitric acid** with the formation of nitrophenic acids (35, *b*), and very gradually decomposed by boiling in solution with formation of sulphates and phenic acid.* Free sulphophenic acid evolves phenic acid when heated to the boiling point of the latter.—The sulphophenates are all soluble in water, and mostly soluble in alcohol.

LIQUID NON-VOLATILE ACID.

38. LACTIC ACID. $HC_3H_5O_3$. *Characterized* by its physical properties (*a*); by the solubility and crystalline form of its salts (*b*); by the extent of its reducing power (*c*).—*Separated* from many acids by the solubility of its lead salt in water, alcohol, and ether (*d*); from glycerin, sugar, etc., by the insolubility of its zinc salt in alcohol (*f*); from tissues, etc., as below (*e*).—*Determined* by saturation with alkali (*g*); by weight of zinc or magnesium salt (*h*).

a. Absolute lactic acid is a colorless, odorless, syrupy liquid, of a very acid taste. Pure, it has the spec. grav. 1.248; when 75 per cent., the spec. grav. 1.212. Not volatile without decomposition; not decomposed by heat below 130° C.; at 145° C. vaporizes dilactic acid, at higher temperature lactide, both of which are converted to lactates by the alkalies.—*Soluble* in all proportions of water, alcohol, and ether; slightly soluble in chloroform. (Glyceric acid, $C_3H_6O_4$, which resembles lactic acid, is insoluble in ether.) Concentrated sulphuric acid mixes with lactic acid without blackening it. Heated on platinum foil, it leaves a slight carbon residue which burns wholly away.

* PRESCOTT : Chem. News, xxvi., 269.

b. The metallic lactates are all *soluble* in water; being mostly sparingly soluble in cold, freely in boiling water. *Calcium* lactate is soluble in 9½ parts (sarcolactate in 12½ parts) of cold water, soluble in alcohol, not in ether. *Barium* lactate is soluble in water and alcohol, insoluble in ether. *Zinc* lactate is soluble in 58 parts of cold, 6 parts of boiling water; insoluble in alcohol (sarcolactate in 6 parts cold water and in 2.2 parts cold alcohol). *Silver* lactate is soluble in water and in hot alcohol. *Lead* lactate is freely soluble in water, sparingly soluble in cold, readily in hot alcohol, slightly soluble in ether. (Glycerate of lead is but slightly soluble in cold water.)

Calcium lactate (saturated with base) *crystallizes* in small white mammillated tufts, seen under the microscope to consist of delicate needles, some of which resemble a bundle of bristles bound midway between the ends. The acid lactate of calcium (supersaturated with acid) forms white hemispheres, compactly made of radiate needles, trimetric. Zinc lactate crystallizes from concentrated solutions in shining crusts, from dilute solutions in four-sided prismatic needles; the crystals, $\mathbf{Zn(C_3H_5O_3)_2.3H_2O}$, lose their water rapidly at 100° C., and the salt decomposes above 210° C. (Zinc Sarcolactate crystallizes in slender needles, $\mathbf{Zn(C_3H_5O_3)_2.2H_2O}$, losing their crystal water very slowly at 100° and giving off empyreumatic vapors below 150°.) Silver lactate crystallizes from neutral solutions, in slender needles, grouped in nodules, quickly blackening in the light.

c. Lactic acid does not *reduce* the alkaline solution of sulphate of copper, but quickly reduces potassium **permanganate** from acid or alkaline solutions.

d. Lactic acid may be *separated* from acids which form insoluble **lead** salts (and other insoluble bodies), according to the general method given at 40, *g*, either in alcoholic or aqueous solution. In a similar manner it is removed from insoluble **barium** salts, as soluble barium lactate, after saturation with carbonate of barium. The barium is then removed from the filtrate by precipitation with sulphuric acid and filtration, and the sulphuric acid is removed from the lactic acid in the last filtrate by repeatedly adding a mixture of 1 part of alcohol and 5 parts of ether and evaporating.

e. Also, the fluid obtained by digestion and expression of *tissues* may be treated with sulphuric acid to fix albuminous matters, filtered, treated with alcohol and five times its weight of ether and again evaporated, filtering when necessary, till the sulphuric acid is removed.

f. A (weighed) quantity of the material containing lactic acid, mixed with substances soluble in alcohol, is saturated in aqueous solution with oxide of **zinc,** the mixture evaporated to dryness, the residue digested in **alcohol** and filtered. The filtrate will contain the substances soluble in alcohol; the residue will contain zinc lactate, soluble in water.

Quantitative.—*g.* In the *acidimetry* of lactic acid, one-tenth equivalent, 9.000 being taken, the required number of cubic centimeters of normal solution of alkali equals the number per cent. of $\mathbf{HC_3H_5O_3}$.

h. Saturating with oxide of zinc or oxide of magnesium, filtering and washing with water, crystallizing or evaporating, and drying at 100° C.:

$$\mathbf{Mg(C_3H_5O_3)_2 : 2HC_3H_5O_3 :: 1 : 0.8911.}$$
$$\mathbf{Zn(C_3H_5O_3)_2 : 2HC_3H_5O_3 :: 1 : 0.7402.}$$

LIQUID VOLATILE ACIDS.

39. FORMIC ACID. $\mathbf{HCHO_2}$. *Identified* by its odor (*a*); by its reducing power upon salts of the noble metals, permanganates, chromates, etc.—the radical $\mathbf{CHO_2}$ being oxidized to $\mathbf{H_2O}$ and $\mathbf{CO_2}$—(*b*); by the color of its ferric salt in solution (*c*); by the odor of its ethyl salt (*d*).—*Separated* from substances less volatile by distillation (*f*); from organic acids in general by the solubility of its lead salt in water (*g*); from acetic acid by the insolubility of its lead salt and its magnesium salt in alcohol (*h*).—*Determined* by acidimetry (*j*), or by oxidation to carbonic anhydride (*k*).

a. The *odor* of formic acid is pungent, irritating, characteristic, slightly acetous, and of an intensity varying greatly with the strength and temperature of its solutions. In contact with the skin, it causes intense irritation.

b. **Nitrate of silver** in concentrated solution gives, with solutions of formates, the white crystalline precipitate of formate of silver, not formed with free formic acid. The precipitate darkens upon standing a short time, and when warmed it is quickly *reduced* to metallic silver. In case the formic acid is free, or the formate in dilute solution, so that formate of silver is not precipitated, the reduction of metallic silver occurs slowly in warm solution. An excess of ammonia retards or prevents the reduction. **Mercuric chloride** in hot solution is gradually reduced by formic acid, more readily by formates, a white precipitate of mercurous chloride forming first, then a dark gray precipitate of metallic mercury. Alkaline chlorides and acetic acid retard or prevent the reduction. Solution of **potassic permanganate** is slowly decolorized at ordinary temperatures, and warm solution of **chromic acid** is gradually turned green, by sufficient formic acid or formates.

Chlorine and bromine oxidize formic acid to carbonic anhydride and hydracid. Nitric acid also decomposes it, likewise peroxide of mercury in boiling solution (removal from acetic acid, see *i*).

c. **Ferric chloride** solution with formates produce a red solution of *ferric formate.*

d. With **alcohol** and sulphuric acid, at a gentle heat, formic acid becomes *formate of ethyl*, $C_2H_5CHO_2$, an ether having a strong, agreeable odor, like that of peach-kernels, and distilling at about 55° C.

e. Strong **sulphuric acid,** at a gentle heat, decomposes $\mathbf{HCHO_2}$ into $\mathbf{H_2O}$ and **CO.** Strong **alkalies** at a gentle heat convert formic acid into *oxalates;* at a higher heat *carbonates* are formed with liberation of carbonic oxide.

f. Absolute formic acid *distils* at 100°; the aqueous solution, 77.5 per cent. of acid, at ordinary atmospheric pressure, boils at 107.1°, and mixtures containing larger or smaller proportions of water are reduced to this per cent. of acid and boiling-point by repeated distillations. A glycerin-bath may be used. Formic and acetic acids are not easily separated by fractional distillation. Dilute sulphuric acid is employed for the production of formic acid from formates.

g. The formates are all *soluble* in water. Plumbic formate requires 40 parts of cold water or a smaller proportion of hot water for solution. Argentic formate is sparingly soluble in cold water, decomposed by hot water (*b*). Mercurous formate is the least soluble salt of this acid, requiring about 500 parts of cold water for solution. It is much more soluble by hot water, in which it decomposes.—In **alcohol,** the formates of lead, magnesium, calcium, and barium are insoluble, the alkaline formates soluble.

Formic acid is *separated* from far the larger number of organic acids by precipitation of the latter as **lead** salts. With free acids, the method given for acetic acid (40, *g*) may be employed, avoiding the use of heat in any part of the operation.

h. Formic acid is separated *from Acetic* acid by saturating with magnesia, or with lead oxide or carbonate, adding much alcohol, filtering and washing with alcohol. In the preparation of formic acid, acetic acid is approximately separated by the crystallization of plumbic formate from water solution containing also plumbic acetate.

i. Formic acid is *removed* from acetic acids, or from other acids not very easily oxidized, by hot digestion with mercuric oxide, until effervescence ceases. **HgO** and $\mathbf{HCHO_2}$ form $\mathbf{H_2O}$ and $\mathbf{CO_2}$ and **Hg.** The filtrate will contain mercuric acetate if acetic acid were present; in fact, the presence of mercury in the filtrate indicates some other acid besides formic. Acids forming insoluble mercury salts may be obtained from the residue, by treatment with hydrosulphuric acid, filtration, and dissipation of the excess of hydrosulphuric acid in the last filtrate.

Quantitative.—*j*. Free formic acid may be determined by the ordinary methods of acidimetry. See 40, *i*, *j*. Or the acid may be saturated with pure carbonate of barium, and the formate of barium precipitated as a sulphate—$BaSO_4$: $2HCHO_2$: : 1 : 0.395.

k. Formic acid is quantitatively separated from acetic acid by precipitation with alcoholic solution of plumbic acetate, washing the precipitate with alcohol. The *formate of lead* may be determined, after oxidation with **chromate** and an acid, as carbonic anhydride. The lead formate, with solution of bichromate of potassium, is placed in an apparatus for determination of carbonic anhydride (from carbonates whose bases form insoluble sulphates), and decomposed by nitric acid, gradually, as the dry gas escapes, in the usual manner. CO_2 : $HCHO_2$: : 1 : 0.956. Or, the carbonic anhydride may be received in an ammoniacal solution of chloride of barium. $BaCO_3$: $HCHO_2$: : 1 : 0.233.

40. ACETIC ACID. $HC_2H_3O_2$. *Identified* by its odor (*a*), by the odor of its ethyl salt (*b*), by the odor arising from the ignition of its salts alone (*c*) or with arsenious acid (*d*), by the color of its ferric salt in solution (*e*), by the free solubility of its lead salt and the sparing solubility of its silver salt (*f*).—*Separated* from less volatile or more volatile substances, by distillation (*h*); from the larger number of acids, by the solubility of its lead salt (*g*).—*Determined* as free acid, or in salts of insoluble bases, by its saturating power (*i*, *j*).

a. Aqueous acetic acid evolves the *odor* of vinegar, which is pungent in proportion to the strength and temperature of the solution. Acetates impart the same odor in a very slight degree.

b. The *acetate of ethyl*, **C_2H_5 $C_2H_3O_2$**, is obtained by warming acetic acid or its salts with **sulphuric acid** and a small proportion of **alcohol.** It is recognized by its pungent and fragrant odor, ethereal, refreshing, and obscurely acetous. It distils at 74°, and may be cleared from acids and from water by contact with dry carbonate of potassium. It is neutral to test-paper, and is soluble in about ten parts of water.

c. When **ignited** in a tube closed at one end, most of the metallic acetates evolve *acetone*, **C_3H_6O**, a vapor of an agreeable odor, readily burning with a white flame. Liquid acetone boils at 56°.

d. If acetates are heated with fixed alkali and **arsenious acid,** the offensive odor of *cacodyl* is observed, **$As_2(C_2H_5)_2$ H_2O.**

e. Solutions of **ferric salts,** with solutions of acetates (not with hydric acetate), form a dark red solution of *ferric acetate*, **$Fe_2(C_2H_3O_2)_6$**, decolorized by strong sulphuric or hydrochloric acid (distinction from Meconate), not decolorized by solution of mercuric chloride (distinction from Sulphocyanate), precipitated as basic acetate by boiling.

f. The metallic acetates are soluble in water, argentic and mercurous acetates being sparingly soluble and forming as crystalline precipitates from concentrated solutions. *Argentic acetate* forms white, fine, scaly crystals, soluble in one hundred parts of cold water and in a smaller proportion of hot water. *Mercurous acetate* forms scaly crystals, sparingly soluble in water, more soluble in dilute acetic acid. The normal and basic acetates of lead are freely soluble in water. In **alcohol,** mercurous and argentic acetates are nearly insoluble, mercuric acetate is slowly decomposed, normal lead acetate freely soluble, basic lead acetates sparingly soluble, the other metallic acetates soluble. Zinc acetate crystallizes in hexagonal plates, very soluble in water.

g. The *solubility of its lead salt* enables (free) acetic acid to be separated from organic acids in general (not lactic, formic, butyric, valeric)—in qualitative or quantitative work—as follows:

Digest the acids in a closed flask at a gentle heat with sufficient oxide of lead, until the mixture is just alkaline to litmus; filter and wash. For complete separation from tartaric acid, or other acid having its lead salt appreciably soluble in water but insoluble in alcohol, the solution should be alcoholic and the washing wholly by alcohol, avoiding the use of much excess of oxide of lead.

Residue (A): plumbic salts of organic acids (excess of oxide of lead).

Filtrate (B): plumbic acetate (basic and not freely soluble in alcohol).

Treat filtrate B, in a long-necked flask, with washed hydrosulphuric acid gas, to complete precipitation; filter and wash with water. Return the filtrate and washings to the flask, insert therein a glass tube and blow air from a bellows through the same until the hydrosulphuric acid is expelled.

Filtrate (C): acetic acid (lactic acid; formic acid; butyric acid; valeric acid).

Treat residue A with washed hydrosulphuric acid gas, until the residue appears wholly black, as seen from beneath the vessel. Filter and wash, and expel the hydrosulphuric acid from the filtrate by a current of air from a bellows, as described above.

Filtrate (D): acids whose lead salts are insoluble in water (or alcohol).

h. Acetic acid boils at 119°. It may be *distilled* from a paraffin or glycerin bath. In distillation from sulphuric acid, the acetic acid is liable to be oxidized to a slight extent, with production of carbonic and sulphurous anhydrides, the latter condensing with the acetic acid. For the acidimetry of the distillate, acetates should be distilled with phosphoric acid (or with hydrochloric acid, and the subsequent determination of the latter by standard solution of silver). Fractional distillation—with or without fractional saturation—may be employed in the separation of acetic acid from other acids more or less volatile than itself. (See, also, Valeric acid, *c.*)

Quantitative.—*i. Free acetic acid,* in absence of other acids, may be determined by *neutralization* with an ascertained quantity of alkali. Different alkalies have been used in **standard solution** for this purpose—as soda, potassa, sodic carbonate, lime dissolved with sugar, ammonio-cupric sulphate. In the solid state, calcined magnesia, crystallized sodic carbonate, and potassic bicarbonate have been employed. Also baric carbonate, the barium dissolved as acetate being then determined as sulphate.

In testing colorless or slightly colored solutions with any of the standard solutions named above, except that of ammonio-cupric sulphate, the *point of saturation is indicated* by litmus; but in case the acetic solution is colored somewhat, a little sulphate of copper may be added, when the neutral point will be indicated by the cloudiness due to the commencing precipitate of hydrate of copper. With the ammonio-cupric standard solution, the solution determined must be very dilute, when saturation will be shown by the turbidity. In the use of calcined magnesia, saturation is indicated by the dissolving of the solid, as well as by the color of litmus.

In the method with carbonate of barium, the acid is saturated with the pure carbonate; the acetate of barium filtered and washed from the excess of the reagent, precipitated by dilute sulphuric acid and weighed as barium sulphate. $BaSO_4 : 2HC_2H_3O_2 :: 1 : 0.515$.

The most convenient *standard* of solutions of alkalies are the "normal solutions," operating upon one-tenth equivalent of the acid—$HC_2H_3O_2$—6.000 grams of the material.

j. The acetic acid producible from *acetates of bases insoluble in water* may be estimated volumetrically, as follows: To a solution of 6.000 grams of the acetate, add normal solution of alkali to complete precipitation, noting the number of cubic centimeters used. Filter and wash till the washings do not affect litmus-paper. To the filtrate and washings, add of a normal solution of acid to the neutral point. The number of cubic centimeters of alkali used, minus the number of cubic centimeters of acid used, expresses the per cent. of acetic acid sought.

41. BUTYRIC ACID. $HC_4H_7O_2$. *Identified* by its odor (*a*); by the odor of its ethyl salt (*b*); by its liquidity, solubilities, and the properties of its salts of lead, barium, and other metals (*c*).—*Separated* from acids having higher or lower boiling points by fractional saturation and distillation (*d*); from many acids by the solubility of its lead salt in water, and from other acids by the solubility of its lead salt in alcohol (*c*. See process *g*, under Acetic acid).—*Determined* by saturation (*e*); by ultimate analysis.

a. The *odor* of butyric acid is like that of rancid butter, but somewhat less offensive, and obscurely acetous, closely resembling that given by slightly rancid butter when heated. It is a strong and persistent odor, not much diminished by dilution of the acid, but increased by warming it. The metallic butyrates are odorless, unless undergoing decomposition.

b. *Butyric ether*—$C_2H_5C_4H_7O_2$—is formed by warming butyric acid or a butyrate with **alcohol** and excess of **sulphuric acid.** It has the odor of pineapples, by which it is readily identified. It rises to the surface of aqueous mixtures, and may be decanted, and purified from acid by addition of chalk and from water by chloride of calcium. It is soluble in all proportions of alcohol and ether, very slightly soluble in water. It distils at 119° C.

c. Absolute butyric acid is a colorless, mobile *liquid*, solidified at very low temperatures, at 15° C. having a specific gravity of .974. It is *soluble* in all proportions of water, alcohol, ether, and wood-spirit. It is not soluble in concentrated solutions of freely soluble salts. The *metallic butyrates* are all soluble in water; plumbic, argentic, and mercurous sparingly soluble; calcic freely soluble in cold water, but sparingly soluble in hot water. Plumbic butyrate is more soluble in **alcohol** than in water; argentic butyrate less soluble in alcohol than in water; baric butyrate very sparingly soluble in alcohol; potassic butyrate freely soluble in alcohol.

Butyrate of lead is formed slowly on adding butyric acid to lead acetate as a heavy liquid which solidifies on standing. Alkaline butyrates, with lead acetate in moderately concentrated solution, give a milky precipitate, which afterward solidifies in a white semi-crystalline mass. A nearly saturated solution of butyrate of lead, left over sulphuric acid, deposits fine, silky needles which are anhydrous. Butyrate of *silver* is formed in shining scales by mixing moderately dilute solutions of nitrate of silver and alkaline butyrate. Butyrate of *copper* forms blue-green monoclinic crystals sparingly soluble in water (see Valeric acid, *b*). Butyrate of zinc crystallizes in shining scales. Butyrate of barium is formed by saturating butyric acid with hydrate of barium, and crystallizes in the cold in long flattened prisms containing 2 aq. Butyrate of *calcium*, obtained in the same way, crystallizes in delicate needles, anhydrous.—Butyrates of lead, barium, calcium, potassium, and some other metals, *rotate* rapidly when dropped in small fragments upon water.

d. Butyric acid distils unchanged at 157° C. Its separation from propionic, acetic, valeric, caproic, and other acids of contiguous boiling points, is best accomplished by fractional saturation and distillation. (43. Also, see Valeric acid, *c.*)

Quantitative.—*e.* Butyric acid has been determined by saturation with (10 parts of dry) **bismuth hydrate**, and precipitation of the butyrate of bismuth with ammonia to obtain the oxide of bismuth, which is dried and weighed. $\mathbf{Bi_2O_3 : 6HC_4H_7O_2 :: 1 : 1.1282}$.

42. VALERIC ACID. $\mathbf{HC_5H_9O_2}$. *Identified* by its odor and taste, the odor of its ethers, and the taste of its alkaline salts (*a*); by its consistence, boiling point, solubilities, and the properties of certain of its metallic salts (*b*).—*Separated* by fractional distillation (*c*); by solubility of certain salts of lead, copper, iron, barium, zinc (*d*).—*Determined* by acidimetry (*e*); approximately, by solubility in water (*f*).

a. The *odor* of valeric acid is that characteristic of dried valerian root and of common valerian oil, in part like that of decayed cheese and also of butyric acid. When not diluted, it has a sour, burning, and disagreeable *taste* and caustic effect.

The alkaline *valerates* have a sweetish taste, with a pungent and alkaline after-taste, and when moist exhale some odor of valeric acid. *Ethyl valerate*, evolved on warming valeric acid or its salts with alcohol and sulphuric acid, has an agreeable, fruity odor. *Amyl valerate*, formed by heating valerianic acid with a very little fusel-oil and sulphuric acid, is characterized by a pleasant apple odor.

b. Absolute valeric acid ("monohydrate") is a transparent and mobile oily liquid, of sp. gr. of .937 at 15° C., boiling at 175° C. With water it forms a definite hydrate—$\mathbf{HC_5H_9O_2 . H_2O}$ ("trihydrate")—an oily liquid of sp. gr. of .950, boiling at 165° C., but gradually dehydrated by distillation, the first distilled portion containing the hydrate mixed with water, after which the absolute acid passes over.—Absolute valeric acid is *soluble* in 30 parts of water at ordinary temperatures; the hydrated acid in 26 parts. It is almost wholly removed from solution by saturation with freely soluble salts, as chloride of calcium or of sodium. It is soluble in all proportions of alcohol, ether, chloroform, and glacial acetic acid.

The *valerates* of the alkali metals are deliquescent and freely *soluble* in water and in alcohol; of the alkaline earth metals, moderately soluble in water and in aqueous alcohol. Aluminum valerate is insoluble. Ferric valerate (basic) insoluble. Zinc valerate is soluble in 90 parts of water, and in 60 parts of alcohol of 80 per cent. Bismuth valerate (basic) insoluble in water; silver valerate, slightly soluble in water; lead valerate (normal) readily soluble in water, (basic) sparingly soluble in water; mercuric valerate, soluble; mercurous, slightly soluble; cupric valerate, moderately soluble.

The lead valerate *crystallizes* in shining needles gathered in hemispherical groups; silver valerate in white, shining plates; copper valerate in green-blue monoclinic prisms; mercury valerate in slender white needles; zinc valerate in snow-white plates of pearly lustre. The sodium and potassium valerates melt at 140° C., and solidify in amorphous cakes, white when pure.

Sodium valerate crystallizes, by spontaneous evaporation in warm and dry air, in cauliflower-shaped masses.—Many of the valerates *rotate* upon the surface of water when dropped in small fragments upon it.

Silver valerate is *precipitated* from solutions of valerates not too dilute in a white curd, turning black in the light.—Solution of acetate of **copper** on agitation with concentrated valeric acid forms anhydrous valerate of copper in oily droplets, which, after five to twenty minutes, crystallize as greenish-blue monoclinic prisms or octahedrons of hydrated cupric valerate, soluble in a moderate quantity of water and in alcohol. (Distinction from *Butyric acid*, which forms in solution of acetate of copper, not very dilute, an immediate precipitate or turbidity of butyrate of copper, bluish-green and finely crystalline in monoclinic prisms—Lorocque and Huraut.)—Valerates are *decomposed* by acetic, tartaric, citric, and malic acids; not by butyric acid.—Valeric acid decolors potassium **permanganate** solution.

c. Valeric acid is easily separated from Butyric acid by *fractional saturation and distillation* of the latter, the butyrate being wholly decomposed at the temperature of the less volatile acid, which remains in the retort as valerate (41, *d*). With Acetic acid, however, the more volatile acid is held by the base in the retort, while valeric acid distils over. In decomposing valerates for distillation of the acid, sulphuric acid may be employed, avoiding a strong excess.

d. Valeric acid is separated from acids which form insoluble **lead** salts by the method given under Acetic acid, *g*. From acids forming soluble salts of **aluminum**, by the insolubility of aluminum valerate.—If a solution of a valerate made slightly alkaline to test-paper is fully decomposed by solution of **ferric** chloride, and after a short time filtered, the filtrate will be red if Acetic acid is present. "A solution of valeric acid in 50 parts of hot water, saturated with hydrated carbonate of **zinc**, yields a liquid which, when filtered and evaporated to 10 parts and cooled, affords white pearly crystals of valerate of zinc. The mother-water, drained from these crystals, should not yield, by further evaporation and cooling, a salt crystallizing in six-sided tables and very soluble in water" (acetate).—Valerate of **barium** is soluble in 2 parts cold water, sparingly soluble in alcohol; Caprylate of barium in 120 parts water, nearly insoluble in alcohol; Caprinate of barium almost insoluble in water.

Quantitative.—*e.* Free valeric acid, in absence of other acids, may be determined by normal volumetric solution of alkali. Weighing 10.2, the number of cub. cent. of alkali solution equals the number per cent. of $\mathbf{HC_5H_9O_2}$; weighing 12., the number of cub. cent. equals the number per cent. of $\mathbf{HC_5H_9O_2.H_2O}$.

f. A weighed quantity of the acid (1 gram in a tared flask) should require not less than 26 times its weight of water at 16° to 18° C. for perfect solution (absence of alcohol, acetic acid, valerates, etc.), and should require not more than 30 times its weight for exact solution (absence of fatty acids, valeral, etc.)—DUFLOS.

43. Formic, Acetic, Butyric, and Valeric acids may be *separated* from each other by **Fractional Saturation** and **Distillation,** as follows: (This method is generally applicable in fractional distillation.)—To one-half of the material to be distilled add enough potassa or soda to neutralize, and then mix with the other half and distil—with a thermometer in the retort or generating flask to show the boiling point—receiving the distillate all together. If the boiling point has been constant, no farther separation can be effected by this method; if not, saturate half the distillate, mix with the remainder, and distil as before. Repeat the fractional saturation with alkali and distillation of the free acid of the receiver until the distillate has a constant boiling point. Now to the several retort residues add excess of dilute sulphuric acid and distil each; if their distillates do not show a constant boiling point, half saturate and distil, in each case, as before, until the boiling points are constant. Again decompose and distil the retort residues, as before, repeating the operations until the whole of the organic acids is obtained in separate distillates, each showing a constant boiling point. The work may be tabulated as follows:

<table>
<tr><th colspan="4">Fractional Saturation and Distillation.</th></tr>
<tr><td colspan="4">Mixture of acids, a, b, c, d, of different boiling points.
Neutralize half the mixed acids and distil.</td></tr>
<tr><td colspan="2">In Retort : salts of c, d.
Saturate with sulphuric acid and distil.
(Boil. point changes.)
Neutralize half and distil.</td><td colspan="2">In Receiver : a, b (boil. point changes).
Neutralize half and distil.</td></tr>
<tr><td>In Retort : salt of d.
Saturate with sulphuric acid and distil.
In Receiver : d.
(Boil.pt. const'nt.)</td><td>In Receiver : c.
(Boil.pt. const'nt.)</td><td>In Retort : salt of b.
Saturate with sulphuric acid and distil.
In Receiver: b.
(Boil. pt. const'nt.)</td><td>In Receiver : a.
(Boil. pt. const.)</td></tr>
</table>

44. VOLATILE FAT ACIDS of the Acetic Series. (Approaching towards these, in their properties, are the volatile acids of the acetic series which do not have a fatty consistence, though commonly termed "volatile fatty acids"—viz., Formic, Acetic [Propylic], Butyric, and Valerianic acids.)

Caproic acid,	$HC_6 H_{11} O_2$,	boil. at 200°C.,	melt. at	9°C.
Œnanthyc acid,	$HC_7 H_{13} O_2$,	" " 218°C.,	" below	20°C.
Caprylic acid,	$HC_8 H_{15} O_2$,	" " 236°C.,	" at	15°C.
Pelargonic acid,	$HC_9 H_{17} O_2$,	" " 260°C.,	" "	10°C.
Capric acid,	$HC_{10} H_{19} O_2$,	" with decom.,	" "	30°C.

Characterized by their pungent and unpleasant odors (when free), by the persistent and fragrant odors of their ethyl ethers, by their liquid and more or less oily consistence at ordinary temperatures and their capability of distillation, by their sparing solubility or insolubility in water and ready solubility in alcohol and in ether, by their acid reaction, by forming with alkalies salts soluble in water.

Separated from each other by **Fractional Crystallization,** as barium salts, as follows: Add to the mixture (aqueous or alcoholic) sufficient potassa to neutralize, and add chloride of barium to decompose. Crystallize, removing the successive crops of crystals:

FROM WATER SOLUTION.	FROM ALCOHOL SOLUTION.
1st crop—baric caprate,	1st crop—baric caprylate,
2d " " pelargonate,	2d " " œnanthate,
3d " " caprylate,	3d " " pelargonate,
4th " " œnanthate,	and caprate,
5th " " caproate.	4th " " caproate.

The aqueous crystal-crops may be washed with hot alcohol—the washings containing the salts, successively, in order the reverse of their crystallization from alcohol. Thus, the third crop of crystals from water, when washed with alcohol, lose first caproate, then caprate and pelargonate, lastly œnanthate, with little loss of caprylate.

Separated, also, by **Fractional Saturation** (43).

FAT ACIDS, LIQUID AND SOLID.

45. NON-VOLATILE FAT ACIDS. *Characterized* by an oily *consistence*, leaving a permanent oil-spot upon paper, and melting at different temperatures, mostly between 14° C. and 80° C.; by *insolubility* in water, upon which they mostly float (in oily drops or layers, liquid if the water is hot); by free solubility in alcohol, the solutions mostly having an acid reaction, and by solubility in ether; by the (soapy) solubility of their alkaline *salts* in water; by the waxy consistence of their lead salts, which melt and do not dissolve in water and have differing solubilities in alcohol and ether; by forming white, milky *precipitates* when their alkaline salts in water solution are treated with salts of metals not alkalies, or with acids, also when (as free acids) their alcohol solutions are diluted with water. The avidity of drying oils for oxygen is a characteristic of their acids. (See Fixed Oils.)

The nine following are some of the more frequently occurring non-volatile fat acids, placed in order of their fusibility:

46. Ricinoleic Acid. $\mathbf{HC_{18}H_{33}O_3}$. Melts at 10° to 6° C. (14° to 21° F.). Yellowish, syrupy, inodorous, of harsh and persistent taste; reddens litmus, and in alcoholic solution decomposes carbonates with effervescence; distils an illy-smelling liquid; its glyceride and all its metallic salts soluble in alcohol, its lead salt soluble in ether. When Castor Oil (ricinoleate of glyceryl) is heated on a sand-bath with a double volume of nitric acid of 25 per cent., until the nitric acid is all removed; the residue saturated with concentrated solution of sodium carbonate—the characteristic odor of œnanthyc acid is obtained.

47. Oleic Acid. $\mathbf{HC_{18}H_{33}O_2}$. Melts at 14° C. (57° F.), soft above 4° C. (39° F.) Colorless, limpid liquid of sp. gr. 0.808, odorless and tasteless, crystallizing from cold alcoholic solution in white needles; reaction neutral, becoming acid on exposure to the air, by which it finally turns brown and rancid. Its lead salt (lead plaster) is insoluble in alcohol, slowly soluble in ether (separation from palmitate, stearate, laurate, etc.) Distilled with nitric acid, all the volatile acids of the acetic series are found in the distillate.

48. Linoleic Acid. $\mathbf{HC_{16}H_{27}O_2}$? Melts at about 18° C. (64° F.); faint yellow, limpid liquid of sp. gr. 0.921, of taste at first mild and afterward harsh; faintly acid to test-paper; oxidizes in the air to a thick, viscid mass, its salts, also, being changed in the air. Most of the linoleates are soluble in alcohol; the lead salt is soluble in ether.

49. Erucic Acid. $C_{22}H_{42}O_2$. Melts at 34° C. (94° F.); crystallizes from alcohol in shining needles; lead salt not soluble in ether (separation from Oleic acid).

50. Lauric Acid. $HC_{12}H_{23}O_2$. Melts at 43° C. (110° F.); solidifies in scales and crystallizes from alcohol in white needles; slightly acid to test-paper; lead salt sparingly soluble in alcohol, insoluble in ether.

51. Myristic Acid. $HC_{14}H_{27}O_2$. Melts at 54° C. (129° F.); crystallizes in shining laminæ; exceptional in being insoluble in ether; the alcoholic solution has an acid reaction; the lead salt is soluble in alcohol, but insoluble in ether; the barium salt nearly insoluble in alcohol.

52. Palmitic Acid. $HC_{16}H_{31}O_2$. Melts at 62° C. (143° F.); colorless, tasteless, odorless, showing an acid reaction; lighter than water; crystallizes, in congealing, in shining scales, from dilute solutions in slender needles; lead salt insoluble in alcohol or cold ether; barium salt sparingly soluble in water or alcohol; calcium salt insoluble in water or ether, slightly soluble in warm alcohol.

53. Stearic Acid. $HC_{18}H_{35}O_2$. Melts at 70° C. (159° F.); inodorous, tasteless, colorless in liquid and white in solid state; crystallizes from alcohol in needles or nacreous scales, having the specific gravity of water; its solutions distinctly acid to test-paper; lead salt insoluble in alcohol or ether, and not wetted by water and fusible at 125° C.; barium salt insoluble in water, alcohol, or ether; magnesium salt insoluble in water, and slightly soluble in cold, more soluble in hot alcohol. [For the fusing-points and modes of solidification of mixtures of Stearic with Lauric, Myristic, and Palmitic acids, as determined by Heintz, see Watts's Dictionary, v., 414.]

54. Cerotic Acid. $H.C_{27}H_{53}O_2$. Melts at 79° C. (174° F.); crystallizes in congealing in small grains, lighter than water; when pure, is capable of distillation; soluble in hot alcohol and in ether, not soluble in chloroform; solutions acid in reaction; lead salt insoluble in alcohol.

55. *The non-volatile Fat Acids* are *separated* from neutral fats by saponification with fixed alkalies, lime, or oxide of lead, in each case effected by hot digestion in presence of water. Sometimes an alcoholic solution of alkaline salt is precipitated by alcoholic acetate of lead (the lead salt being insoluble in alcohol); in other cases, an alcoholic solution of lead salt is precipitated by alcoholic acetate of barium or of magnesium (such being the solubilities of the respective salts). Then the purified salt is decomposed in water with dilute acid.

As in manufacturing operations, the neutral fats may be decomposed by superheated steam, with separation of the fat acids together.

56. The fat acids are in some cases *separated from each other* by **fractional fusion** of their glycerides, with pressure. The melting point of the glycerides (the neutral fats), is given in 59. The melting point of a mixture of free fatty acids is generally much below the mean melting point of its constituents, as shown by the tables of Heintz mentioned in 54, and hence in many cases no separation can be accomplished by fractional fusion. Thus, free stearic acid can be freed from oleic but not from lauric, myristic, or palmitic acid, by this process.

57. The use of solvents in separation—of the free acids or of their salts—is indicated to some extent by the statements of solubilities, given in this work or elsewhere, and more particularly by the various *methods of preparation* of the acids in question, as found in Watts' Dictionary, Miller's Organic, Gmelin's Handbook, and in original reports.

Free fatty acids are *separated from neutral Fat oils* (not from castor oil), for commercial determinations, by extracting the oil with one or two volumes of 90 per cent. **alcohol.** The acid is then determined volumetrically with soda solution.*

Also, by **alkaline carbonates,** which at ordinary temperatures saponify with fat acids but not with fats. Prepare a solution of

* Burstyn: *Zeitschr. Anal. Chem.*, xi., 283.

10 grams crystallized sodic carbonate, 1 gram sodic bicarbonate, and 30 c. c. water. Agitate, in a test-tube, equal volumes of this solution and of the oil, and set aside at ordinary temperatures. In absence of fat acids, the two liquids separate, more or less turbid; if fat acids are present, an emulsion is formed (from which a cream rises after some time). Old fat oils usually contain traces of fat-acids, scarcely indicated in this test.

58. For the *quantitative* determination of free fat-acid in mixture with neutral fats, digest 10.0 grams of the oil with 2.5 grams of pulverized sodic bicarbonate and 25 drops of water, on a water-bath, with trituration, for an hour. When cold, extract with petroleum naphtha, stirring; evaporate the naphtha, and weigh the neutral fat so separated. Benzole is not applicable in this separation.

NEUTRAL SUBSTANCES, LIQUID OR FUSIBLE.

59. FIXED OILS. Fats or Fat-oils. Glycerides of the non-volatile fat acids. (The following list includes those of most frequent occurrence in commerce.)

a. Liquid at ordinary Temperatures.

aa. Drying Oils (not forming Elaidin).

	Spec. grav.	*Congeal. pt.*	
Hemp-seed, . .	0.926	-25° C., -13° F.	Greenish when fresh, afterward brownish-yellow; unpleasant odor and insipid taste.
Grape-seed, . .	.918	-13° C., 9° F.	Yellow to brownish; nearly odorless, of mild taste.
Linseed, . . .	.934	-27° C., -17° F.	Gold-yellow to brownish; strong odor and taste.
Poppy-seed, . .	.924	-18° C., 0° F.	Straw-yellow; limpid; feebly pleasant odor and taste.
Walnut, . . .	.925	-18° C., 0° F.	Slightly greenish or yellowish; thick; nearly odorless, of mild nutty taste.

bb. Oils drying to a slight extent and slowly forming a little Elaidin.

	Spec. grav.	*Congeal. pt.*		
Beechnut, . .	.920	−18° C.,	0° F.	Yellowish ; nearly odorless and of a mild taste.
Cotton-seed, . .	.925	1° C.,	34° F.	Yellow or brownish-yellow to colorless ; of mild taste.
Croton, . . .	.942	——	——	Clear, slightly yellow ; of a taste at first mild and then burning and persistent ; causes pustules on the skin.
Sesame, . . .	.921	0° C.,	32° F.	Yellow ; of mild odor and taste.
Sunflower, . .	.924	−15° C.,	5° F.	Yellowish ; limpid ; nearly odorless and tasteless.

cc. Oils not drying, but not forming Elaidin.

Cod-liver, . . .	.930	below	14° F.	Clear yellow to red-brown ; acid reaction ; characteristic fishy odor and taste.
Whale, ...	.925	0° C.,	32° F.	Brownish ; of characteristic disagreeable odor and taste.

dd. Non-drying Oils, forming Elaidin.

Almond, . . .	.918	−20° C.,	4° F.	Clear straw-yellow ; limpid ; inodorous, of a bland, sweetish taste.
Castor, ...	.963	−15° C.,	5° F.	Colorless or slight yellow ; syrupy ; odorless, of mild taste with acrid after-taste. (Sometimes classed among the slightly drying oils.)
Colza, ...	.914	−6° C.,	21° F.	Clear, yellowish ; limpid.
Hazel-nut, . .	.920	−19° C.,	−2° F.	
Lard, ...	.915	10° to	0° C.	Colorless or nearly so ; slight odor of lard.
Mustard (black),	.915	15° C.,	5° F.	Yellowish ; odorless, of mild characteristic taste.
Mustard (white),	.913	(not solidified).		Similar to the above.
Neatsfoot, . .	—	(below 0° C.)		Yellowish ; inodorous, of a bland taste.
Olive, ...	.916	5° C. to 2° C.		Greenish or yellowish ; thick flowing ; of slight pleasant or no odor and mild sweetish taste.
Sperm, . . .	.875	——	——	Limpid ; nearly odorless.
Rape-seed, . .	.914	−6° C.,	21° F.	Clear, yellowish ; disagreeable odor and taste.

b. Solid at ordinary Temperatures.

	Melting.		*Melting.*
Butter, ...	27° to 30° C.	Tallow, Mutton, . .	46° to 50° C.
Cacao butter, . . .	25° to 30° C.	Spermaceti, ...	38° to 47° C.
Lard, ...	28° to 32° C.	Wax, Yellow (Bees'),	60° to 63° C.
Tallow, Beef, . . .	36° to 40° C.	Wax, White, . . .	65° to 69° C.

60. Fixed or Fat Oils are *characterized* by their oily consistence and the physical properties stated above; by their solubilities (*a*) and cohesion-figures on water (*b*); by a neutral reaction; by saponification—forming soapy-soluble compounds with alkalies and waxy compounds with lead oxide (*c*); by giving reactions for glycerin (*d*); by the precipitates obtained from their soap-solutions (*e*); by either oxidizing to a viscid mass in the air (*f*), or forming elaidin with nitric acid (*g*); by their sensible reactions with special reagents (*h*).

a. Insoluble in water, upon the surface of which they float. Mostly insoluble or slightly soluble in alcohol; but Castor oil is soluble in all proportions of absolute alcohol, Spermaceti in 7 parts of boiling absolute alcohol, and Wax partly soluble in alcohol. Soluble in Ether and in Benzole, less freely soluble in petroleum naphtha and in chloroform. (Solid fats are slightly soluble in petroleum naphtha; liquid fats moderately soluble.) Miscible with volatile oils, not with glycerin.

By violent agitation with water, fixed oils form milky mixtures (*emulsions*) from which the oil quickly separates in drops; by agitation or trituration with water mucilages of gums, albumen, gelatin, sugar, and of salts, more perfect mixtures are formed, from which the oil slowly separates as a cream, still containing a little water solution and holding the oil in its characteristic microscopic spheres.

b. If a drop of oil is let fall upon a still surface of perfectly pure water, the oil spreads in a film which breaks into a figure (*cohesion figure*) or succession of figures, characteristic of each oil—fixed oils not being distinguished from volatile oils otherwise than from each other. The formation of these figures constitutes a practicable means of identifying the separate oils, and even to some extent of recognizing them when in mixture.*

c. Saponification is effected in presence of water by digesting with excess of alkali for some time, or with oxide of lead at 100° C. for a longer time. The alkali-soaps dissolve in water, the solution being slightly milky, and becoming more turbid on dilution, and dissolve in alcohol, but mostly refuse to dissolve in ether. The lead soaps of some of the fat acids are soluble in ether; they are fusible, waxy compounds. See Non-volatile Fat Acids (45).

d. The glycerin formed in saponification with oxide of lead or with lime, as above, when separated by the concentration of the clear water solution, renders evidence of its identity, by means of tests given under the head of glycerin (66).

e. The alkali-soap solutions give white *precipitates* with solutions of salts of metals not alkaline, and with acids give white precipitates soluble in alcohol.

f. THE DRYING OILS are recognized by not forming elaidin, when treated as stated in the next paragraph; by drying to a resinous film when spread and exposed to the air, and by inducing elevation of temperature, and in many instances ignition, when diffused through a mass of wool or other porous material and exposed to the air.

g. THE NON-DRYING OR ELAIDIN-FORMING OILS are known by reaction with peroxide of nitrogen. A concentrated solution of **mercuric nitrate**, or nitric acid of brown-red color, may be used. For the reactions given in the following table,† a little of the oil is taken in a test-tube, an equal volume of **nitric acid** of about 25 per cent. is added, the test-tube briefly shaken, a strip of **copper** turnings added, and the whole set aside at ordinary warm temperature, to be examined each quarter of an hour.

* TOMLINSON, MOFFAT: *Chem. News*, 1869. CRANE: *Am. Jour. Phar.*, 1874, Sept.

† HAGER'S *Untersuchungen*, ii., 506.

Tests for Elaidin.

Oils.	Result after ¼ to 2 hours.	Result after standing 8 hours to 2 days.
Non-drying Oils.		
Almond:		
From sweet almonds.	White; cloudy.	White or whitish mass, granular after shaking. Appears homogeneous after 8 to 12 hrs.
From bitter almonds.	White or yellowish-white; more or less turbid.	Yellowish; only partly solidified, with a surface layer of semi-liquid, transparent oil.
Bone, ...	Whitish-yellow.	Nearly all solid; a clear-yellow oil layer, with a whitish crystalline finely granular precipitate.
Castor, ...	Whitish.	Whitish; solidifying after 8 hours or earlier.
Lard, ...	Whitish-yellow.	Whitish or yellow-white; somewhat granular, with transparent spots; rigid; sometimes with a half-liquid surface layer.
Olive:		
Green, . . .	White cloudiness, often modified by color of the oil.	White or yellowish-brown-white solid, made granular by shaking. The mass appears uniform after 4 to 8 hrs.
Yellow, . . .	White or whitish cloudiness.	White or yellowish-white mass, granular after shaking. After 4 to 8 hours the surface appears nearly uniform.
Rape-seed:		
Crude, . . .	Yellow-brown to red-brown.	Reddish-yellow; solidifying after 16 to 24 hours and becoming brownish-yellow; somewhat granular after shaking, the granules enclosed in an oil layer.
Refined, . . .	Whitish-yel'w to br'wn-yellow.	Reddish-yellow; solidifying after 16 to 24 hours and becoming yellow; made granular or pasty by shaking, the granules oil-coated.
Oils drying imperfectly.		
Beech-nut, . .	Yellow or reddish-yellow.	Syrupy; nearly clear; after 2 days a just perceptible separation of elaidin.
Cotton-seed, . .	Reddish-yel'w or br'wnish.	Pasty or syrupy; frequently showing a clear brown-yellow oil layer of one-half to one-third the mixture. Appearance of a precipt. after 1 day.
Sesame, . . .	Red to dark red.	Blackish yellow-brown or red-brown; opaque; pasty. After 1 day a transparent oil layer sometimes appears at bottom or top.

Tests for Elaidin—*Continued.*

Oils.	Result after ½ to 2 hours.	Result after standing 8 hours to 2 days.
Sunflower, . .	Yellowish or faintly reddish.	Brownish yellow. Pasty after 1 day.
Drying Oils.		
Hemp-seed, . .	Green.	Yellow; liquid, nearly or quite clear.
Linseed, . . .	Scarcely changed.	Reddish-brown; liquid and transparent.
Poppy-seed, . .	Scarcely changed.	Reddish yellow-brown or reddish yellow; transparent, liquid.
Walnut, . . .	Scarcely changed.	Yellow; clear liquid.
Non-drying Oils not forming Elaidin.		
Cod-liver, . .	Not changed.	Yellowish-red or reddish-br'n; liquid and transparent.
Croton, . . .	Unchanged or made clearer.	Thick liquid; clear.

If *drying are mixed with non-drying oils,* the latter are easily detected; the former only with greater care. The elaidin mass remains partly liquid, or an oily layer separates from it. To detect an intermixture of drying oil, proceed (with a weighed quantity) as above directed, leaving the mixture about two days, then set it aside at 22° to 25° C. (72° to 77° F.) for 12 hours, and return it, without agitation, to ordinary temperature. The drying oil will now be found more or less perfectly separated from the elaidin. (For finding the proportion of the drying oil, bring a tared roll of blotting paper into contact with the mass, while the temperature is 8° to 10° C. [46° to 50° F.] The increase in the weight of the paper or the loss in the weight of the mixture approximates the weight of the drying oil.)

h. Sulphuric acid, Nitric acid, Phosphoric acid, caustic Alkali, and Nitrate of Silver are the chief of the special reagents for color-tests of fixed oils.

The *test by sulphuric acid* is applied as follows: About 8 drops of the oil are placed in a watch-glass over white paper, and then 2 drops of sulphuric acid of specific gravity of 1.820 to 1.830 (not more concentrated) are dropped near the edge of the glass so as to flow upon the oil. The results are tabulated below;

Sulphuric Acid Test.

OIL.	WITHOUT STIRRING.	AFTER A LITTLE STIRRING.
Almond, . .	Clear; yellow.	Blackish-yellow.
Castor, . .	A tinge of pale brown.	Faintly blackish brown.
Cod-liver, . .	First violet, then red.	Brown-red, with violet rim, finally dark brown.
Lard, . . .	Brownish yellow.	Brown.
Linseed, . .	Brown-red.	Black-brown.
Olive, . . .	Yellow.	Blackish-brown.
Poppy-seed, .	Yellow.	Brownish olive-green.
Rape-seed: Crude	Greenish-blue.	Greenish-blue.
" Refined	Brownish-yellow.	
Whale, . .	Red, afterward violet.	Brown-red to dark brown.
Cotton seed	Orange yellow	Brown-red.

61. *Preparation and Application of Reagents for Identification of Fixed Oils,* according to the following table [CALVERT]:*

(1) *Soda solution.* Specific gravity 1.33. 4 parts of dry soda in 6 to 7 parts of water. One volume of this solution agitated with 4 to 5 volumes of the oil and heated to boiling. (The drying oils, so treated, form soft soaps; the non-drying oils, mostly hard soaps.)

(2) *Sulphuric acid* of spec. grav. 1.475. Mixture of 10 parts of the acid of spec. grav. 1.840 and 7 parts of distilled water. One volume is mixed with 5 volumes of the oil and set aside for ten minutes.

(3) Sulphuric acid of spec. grav. 1.530. Mixture of 10 parts of acid of 1.840 and 6 parts of water. Mix 1 volume with 5 volumes of the oil and set aside for five minutes.

* *Phar. Jour.*, xiii., 356.

(4) Sulphuric acid of spec. grav. 1.635. Mixture of 10 parts of acid of 1.840 and 4 parts of water. Applied like reagent (3).

(5) *Nitric acid* of spec. grav. 1.180. Mix 1 volume with 5 volumes of the oil and set aside five minutes.

(6) Nitric acid of spec. grav. 1.220. Applied as directed for reagent (5).

(7) Nitric acid of spec. grav. 1.330. Applied as directed for (5) and results noted; then an excess of the soda solution (1) is added and results noted again.

(8) *Phosphoric acid* of syrupy consistence.

(9) *Sulphuric acid* of spec. grav. 1.840 with equal measure of *Nitric acid* of spec. grav. 1.330. One volume of the mixed acids for 5 volumes of oil.

(10) *Nitrohydrochloric acid*—from 1 volume of nitric acid of spec. grav. 1.330 and 25 volumes of hydrochloric acid. One volume of the mixture to 5 volumes of oil, noting the result. Then add excess of Soda solution (1), and again note the result.

Oils.	(1) Soda.	(2) Sulphuric Acid.	(3)	(4)	(5) Nitric Acid.	(6)	(7) Nitric Acid, *then* Soda.		(8) Phos. Acid.	(9) Sul.&Nit. Acids.	(10) Aqua Regia, *then* Soda.	
	1.330	1.475	1.530	1.635	1.180	1.220	1.330.	1.330.				
Castor.	White.	. . .	Dirty white.	. .	. . .	. . .	. . .	White; fibrous.	. . .	Brownish-red.	. . .	Pale rose; fibrous.
Cocoa-nut.	Thick; white.	. . .	Dirty white.	Light brown.	. . .	. . .	. . .	Fibrous.	. . .	Orange-white.	. . .	White; fibrous.
Cod-liver.	Dark red.	Purple.	Purple.	Deep brown.	. . .	. . .	Red.	Fluid.	D'rk red.	Dark b'wn.	Yellow.	Orange-yel.; fluid.
Hemp-se'd	Bro'n-yel.; thick.	Bright green.	Bright green.	Bright green.	Dirty gr'n.	Greenish dirty b'wn.	Greenish dirty b'wn.	Lig't b'wn; fibrous.	Green.	Green; then bla'k.	Green.	L'ht b'wn; fibrous.
Lard.	Pinkish-white.	Dirty white.	Dirty white.	Light brown.	. . .	. . .	Faint yel.	Fluid.	. . .	Brown.	. . .	Pink; fluid
Linseed.	Yellow; fluid.	Green.	Dirty green.	Green.	Yellow.	Yellow.	Gr'n; then brown.	Yellow; fluid.	Br'wnish yel.-gr'n.	Green; then bla'k.	Green-yel.	Orange; fluid.
Neatsfoot	Dirty yel.-white.	Yellow tinge.	Br'wnish dirty w'e	Brown.	Light yel.	Light yel.	Lig't b'wn.	Fibrous.	. . .	Dark b'wn.	Slight yel.	Bro'n-yel.; fibrous.
Olive.	Slight yel.	Green tinge.	Greenish white.	Light green.	Greenish.	Greenish.	Greenish.	White; fluid.	Slight green.	Orange-yellow.	. . .	White; fluid.
Poppy-se'd	Dirty yel.-white.	. . .	Dirty white.	. . .	. . .	Ora'ge-yel.	Red.	Light red; fluid.	. . .	Slight yel.	. . .	Intense rose; fluid.
Rape-seed.	Dirty yel.-white.	. . .	Pink.	. . .	. . .	. . .	. . .	White; fluid.	. . .	Dark b'wn.	. . .	Yel. white; fibrous.
Seal.	Dark red.	Light red.	Red.	Bright brown.	Pink.	Light red.	Red.	Fluid.	D'rk red.	Dark b'wn.	Slight yel.	Orange-yel.; fluid.
Sesame.	Dirty yel.-white.	Green tinge.	Greenish dirty w'e	. . .	Ora'ge-yel.	Red.	Dark red.	Red; fluid.	. . .	Gr'n; then bright red.	Yellow.	Orange; fluid.
Sperm.	Dark red.	Light red.	Red.	Bright brown.	Slight yel.	Light yel.	Red.	Fluid.	D'rk red.	Dark b'wn.	Slight yel.	Orange-yel.; fluid.

62. Tests with **Nitrate of Silver.** A two per cent. alcoholic solution of nitrate of silver is prepared: 0.5 gram of crystallized silver nitrate being dissolved in 1.0 gram water and mixed with 25 c. c. of absolute alcohol. Now place 6 or 7 c. c. of the oil in a test-tube about 12 millimetres (0.47 inch) thick, add 2 or 3 c. c. of the silver solution, shake briskly to form a milky mixture, heat, without bringing the tube in contact with the flame, to boiling for a quarter of a minute, and set aside for an hour or two. A *reduction of silver*, with *darkening of the oil layer* to brown, red-brown, or black, results from this test with

Almond oil, from bitter almonds—colored after several hours.
Bone oil—brown to black.
Cotton-seed oil—brown to black.
Lard oil.
Linseed oil—darkens, red-brown.
Rape-seed oil—brown-red.

With the following oils there is no change:

Almond oil,
from sweet almonds,
Beech-nut oil,
Castor oil,
Cod-liver oil,
Hemp-seed oil,
Olive oil,
Sesame oil.

63. **Special examination of Butter.**—*Separation of fats* from non-fatty substances by melting (*a*), by benzole (*b*). *Identification of butyrin*, etc., by etherization after saponification (see Butyric acid, 41, *b*). *Distinction from* (mixtures of) *lard* by treatment with sulphuric acid (*c*), by treatment with ether at 18.5° C. (*d*), or with petroleum naphtha at 10° to 15° C. (*e*); from foreign color by borax solution (*f*).

a. About 10 grams of the butter are melted in a large test-tube, by insertion in water of 50° to 60° C. (122° to 140° F.) for about an hour. The fats separate from a subsident layer of water, casein, salt, lactose, (foreign colors). The volume of the latter may be approximately ascertained by linear measurement on the tube. The fat layer from unsophisticated butter is clear, and has a yellow color of a tint somewhat deeper than that of the butter; while the bottom layer is white, or at most but yellowish-white. (The bottom layer may be $\frac{1}{5}$ at most; from good table butter should not be over $\frac{1}{7}$.)

b. In a large and strong test-tube place 5 grams of the butter, melt by dipping in water at 60° C. (140° F.), add fully an equal volume of **benzole**, cork securely, agitate, and leave at about 40° C. (104° F.) for an hour. The sediment separates more sharply and of thicker consistence than in *a.*—The sediment may be washed with benzole on a filter, and analyzed chemically and microscopically (see *c*, *d*, *e*).

For the separation by benzole, a graduated tube may be used, as follows (HOORN): A glass tube is prepared, 20 centimetres (8 inches) long, its upper two-thirds having a diameter of 2 centimetres (0.8 inch), its lower third narrowed and graduated to tenths c. c., and its lower end closed. In this tube 10 grams of butter are placed, melted by dipping in warm water; 30 c. c. of benzole are added, the contents thoroughly intermixed, and the tube set aside. After thirty to forty minutes, the benzole and fat will have separated from the water layer below—the amount of which may be read off.

c. Take 2 c. c. of the fats separated by melting as in *a*, bring the temperature to about 30° C. (86° F.), add about 3 c. c. of concentrated **sulphuric acid**, and agitate gently to a complete mixture. With butter alone, the liquid remains of a yellow becoming yellow-red color, clear and translucent, not darkening at ordinary temperature, and after half an hour becoming gelatinous and rather less translucent. If Tallow or Lard is present, the mixture after a short time becomes darker, by aid of the heat generated by the acid, so that after half an hour it is dark brown-red or brown-black.

d. The butter is melted over the water-bath, and after standing the liquid fat is removed from the subsident layer. This fat is mixed in an evaporating dish with four or five times its bulk of hot water and left two or three hours. The solidified fat is dried on blotting-paper, introduced into a wide-necked flask, and covered with ether at a temperature of 18.5° C. (65.3° F.) If the butter was pure, the fat fully dissolves to a clear, lemon-yellow liquid. If the butter contained Lard, the fat is in some part insoluble in ether at this temperature, and the mixture is left milky or thick, depositing a (finely granular) sediment on standing. Tallow of beef or mutton gives the same results, the sediment being coarser than in the case of lard. The temperature of the ether is the important condition in this test, and it must not be disturbed by contact with the hand. (HORSLEY.)*

By special apparatus, closer observations are made with this test (BALLARD) as follows: Select a test-tube 11 or 12 centimetres (4 or 5 inches) long and about 2.5 centimetres (nearly 1 inch) wide; and prepare a section of glass tubing of 1.3 to 1.6 centimetres (a little over ½ inch) diameter, and 4 to 5 centimetres (1½ to 2 inches) long, each end being slightly rimmed outward, and the one (lower) end bound over with a bit of thin canvas. Weigh the little tube, with the covered end, and place in it 1.5 grams of the butter to be tested, and in the test-tube 5 c. c. of ether. Attach a thread to the small glass tube and let it down into the ether, then close the test-tube with a cork, so as to hold the thread, and bind the cork over with leather. Immerse the test-tube in water at exactly 18.5° C. (65.3° F.) and leave it an hour at this temperature. The cap is now removed, the small tube drawn up out of the ether by the thread (without removing the cork), and left at same temperature to drain. The small tube is now taken out, and while the top is closed by the finger the liquid is absorbed as far as possible by blotting-paper, the tube exposed to the air till free from ether odor, and weighed.

	With 5 *c.c. ether.*	*With* 10 *c.c. ether.*
From 1.5 grams pure butter, remained insoluble, ...	0.18 grms.	0.14 grms.
From 1.5 grams beef tallow, ...	0.945 "	
From 1.5 grams lard, ...	0.9 "	

* Farther, see Chem. News, Sept. 11, 1874, p. 135.

From equal parts tallow and butter, .	0.6 grms.	
From $\frac{1}{4}$ tallow and $\frac{3}{4}$ butter, . . .	0.3 "	0.8 grms.
From $\frac{1}{4}$ lard and $\frac{3}{4}$ butter, ...	0.15 "	0.8 "
From $\frac{1}{6}$ lard and $\frac{5}{6}$ butter, ...		0.67 "

e. The fat of butter, separated according to *a*, is treated with 7 parts of petroleum naphtha at a low temperature—10° to 15° C.—when the fat of butter dissolves, and tallow, or lard if over 10 per cent., remains in sediment.

f. Boil gently, in a test-tube, 2 grams butter with 5 c.c. of cold-saturated solution of borax, and set aside to cool and subside. Butter not sophisticated leaves the borax solution nearly or quite colorless (with white turbidity); artificially colored butter leaves the borax solution more or less brown.

64. THE FATS ARE DETERMINED *in Milk*—by separation with ether, from the milk (*a*), from the residue (*b*); by opacity of the milk (*c*); approximately and for comparison by the volume of cream (*d*).

a. To 20 c.c. of milk add an equal volume of 10 per cent. solution of potassa (to hold the casein in solution), in a cylinder, and repeatedly extract with ether. [Benzole & Petroleum Naphtha sometimes used.] Dry the ether residue at 110° C. [Farther, see Phar. Jour., 1874, Sept. 5, p. 188; also Wanklyn's Milk Analysis, New York, 1874, p. 24.]

b. Evaporate 10 grams of milk—with 5 grams (fresh dried) charcoal powder or 15 grams (just ignited) ferric oxide or baric sulphate—at 100° C., till the weight is constant (total solids). Extract the residue, while dry (it being very hygroscopic), with ether, and dry the ether residue at 110° C.

c. Use of Vogel's Lactoscope. A test-glass made of two semi-circular glass plates, set parallel and *exactly* 0.5 centimetre apart, to hold a liquid between. In a mixing glass, to 100 c.c. of water, add milk from a pipette, drop by drop, until the diluted milk, when examined in the test-glass, cuts off the light of a candle placed at 10 to 20 inches distance from the glass (the examination being made in a dark room). Dividing 23.2 by the number of c.c. of milk required (to obstruct the light), then adding to the quotient 0.23, the sum is the per cent. of fats in the milk.

d. The milk is set in a (wide) graduated cylinder until the cream has fully separated, when its volume can be read off. (The volume per cent. of cream in cow's milk varies from 5 to 14.)

For Quantitative Analysis of Milk, see, farther, 167 and 168.

65. Fixed Oils are separated *from Volatile Oils* by extraction of the latter with alcohol (not applicable in case of castor oil, which is soluble in all proportions of absolute alcohol or in 4 or 5 parts of 90 per cent. alcohol).—They are also removed from volatile oils by saponification with alkalies and water.—They are *separated from substances soluble in water* by action of that solvent; from various solids by digestion with ether, bisulphide of carbon, benzole or petroleum naphtha; from *emulsions* by spontaneous separation in cream and melting of the latter, or by ether or benzole, with addition of alkali if necessary to prevent coagulation of the emulsifying substance.

Fixed oils are in many cases separated from each other by fractional fusion, according to differences of melting point as stated in the list, this means of separation being subject to the same limitations mentioned as pertaining to Fat Acids (56).—Drying oils are separated from non-drying by transformation of the latter into elaidin, as already directed.

66. GLYCERIN. $C_3H_5(HO)_3$. *Characterized* by its physical properties (*a*); by the products of its decomposition when heated (*b*); by the limits of its reducing power and its interference with precipitation of metallic bases (*c*).—*Separated* from solids by its liquidity at low temperatures; from volatile bodies by their distillation; from sugar, gum, or gelatin by certain mixed solvents (*a*). Its proportion in mixture with water is determined from specific gravity, by use of a table.

a. A colorless, syrupy liquid, of specific gravity 1.267 at 15° C., not congealed at 18° C. (0° F.), mostly separating as a liquid during the freezing of its water mixtures; distilling very slowly with steam at 100° C., slowly giving off vapor with partial decomposition at 120° C. (248° F.), boiling with decomposition of the most part at 290° C. (554° F.) Odorless, and of a pure, sweet taste, and neutral reaction. Soluble in all proportions of water and of alcohol; only very slightly soluble in ether, excess of which separates alcohol from it; not soluble in chloroform; soluble in a mixture of 2 volumes of absolute alcohol and 1 volume of ether (separation from Sugar, Gum, Gelatin, etc.); soluble in a mixture of equal weights of chloroform and alcohol (separation from Sugar, Dextrin, Gum, Extractives—the mixture not acid); not soluble in benzole, bisulphide of carbon, petroleum naphtha, or fixed oils. It dissolves nearly all organic substances soluble in water and many of those soluble in alcohol, most salts of alkaloids, and all deliquescent salts of metals. It dissolves baryta, strontia, and lime, with combination, and potassa and soda with gradual decomposition. It holds salts of iron and copper in solution not precipitated by alkalies. It dissolves one-fifth per cent., each, of sulphur and phosphorus, 20 per cent. of arsenious acid, 10 per cent. of benzoic acid, 15 per cent. of tannic acid (as a waxy solid melting at the temperature of the body), and dissolves and preserves hydrosulphuric acid. It strongly absorbs water from the air. It dissolves iodine freely without decomposition, bromine sparingly with gradual decomposition, and is changed by chlorine and by nitric acid. It combines with strong sulphuric acid, without color or effervescence, as the instable glycerosulphuric acid.

b. At its boiling point, as above, glycerin evolves *suffocating vapors* of acrolein, etc., which vapors may be condensed by ice to a liquid, chiefly acrolein, with some acrylic acid, acetic acid, etc. Acrolein is a very acrid body, boiling at 51° C. (124° F.), soluble in 40 parts of water. With acid sulphate of potassium, glycerin evolves acrolein at lower temperature. Decomposed and vaporized in an evaporating dish over a lamp or sand-bath, only a slight carbon residue remains, staining the dish (distinction from mixture of Sugar, Gums, etc., which leave a puffy carbon residue).

c. Glycerin does not *reduce* hot alkaline sulphate of **copper** solution (distinction from Sugars, etc.); does not reduce nitrate of **silver**, even on addition of ammonia, if dilute and not heated (distinction from admixtures of Formic acid and certain empyreumatic matters), but on boiling it does reduce ammoniacal nitrate of silver solution. (Acrolein, Butyric acid, etc., form white precipitates with silver nitrate, blackening on standing or heating.) At a boiling heat glycerin liberates **iodine** from iodic acid.

d. As concentrated by evaporation in the air from a water-bath, glycerin retains about 5 per cent. of water. The U. S. Pharmacopœia requires spec. grav. 1.25; the German Pharmacopœia spec. grav. 1.23 to 1.25.

GLYCERIN P. C.	SP. GR.	FREEZING.	GLYCERIN P. C.	SP. GR.	FREEZING.
10	1.024	1° C.	60	1.159	below 35° C.
20	1.051	2.5° C.	70	1.179	
34	1.075	6° C.	80	1.120	
40	1.105	17.5° C.	90	1.232	
50	1.127	31.34° C.	94	1.241	

67. SOAPS. *Alkali* salts of Fatty acids (and of Resin acids).—*Characterized* by their peculiar touch and consistence, solid or gelatinous; if solid, by melting or softening when warmed, and more readily if retaining more water; by dissolving in water to a slightly cloudy solution, viscid if concentrated, and made more turbid by dilution, also dissolving in alcohol (the solution being often turbid from fats, alkaline carbonates, or other impurities); by their aqueous solutions being precipitated by salts of metals not alkalies, or by acetic or stronger acids. In the last-named precipitation, the fatty acid will separate as a cream, and may be examined as provided under head of Fat Acids, and the base in solution determined by inorganic analysis.

Soap solutions are precipitated, physically, by common salt, potassa soaps becoming soda soaps by double decomposition. The oleate of potassa is (sparingly) soluble in ether; otherwise the alkaline oleates, stearates, and palmitates are slightly or not at all soluble in ether.

Quantitative.—*a.* In *determining the water* of soaps by direct evaporation, the fine shavings are exposed at first to a temperature of 40° to 50° C., which is after some time increased gradually, so as not to fuse, to 100° C., the latter continued until there is no longer a loss of weight. Stearates so treated still retain about 2 per cent. water.—A more satisfactory determination of the water is effected by dissolving 1 to 2 grams soap in the least sufficient quantity of strong alcohol, adding a weighed quantity of fine sand, just dried, then evaporating, with trituration, and drying at 110° C.—The water is also estimated as remainder after finding the fat acids, bases (combined, free, and carbonated), glycerin, resin, salts, color-substances, and foreign matter.

b. The *amount of absolute soap is determined* from the fat acids approximately (GRÆGER) as calcium precipitate after solution in alcohol. Ten grams of the soap, in fine shavings, are dissolved in 90 c.c. of 90 per cent. alcohol, the solution made up by addition of alcohol to 100 c.c., left to subside, and 10 c.c. of the clear solution are taken out, diluted with water, and precipitated with calcic chloride. The precipitate is gathered in a tared filter, washed, dried at 100° C., and weighed. 100 parts of this precipitate indicate 101.5 parts of anhydrous soda soap.

c. The *fat acids also are determined gravimetrically*, by weight as free acids, by intermixture with beeswax (HAGER), as follows: 10 grams of the soap are dissolved by warming in an evaporating dish in about 50 c.c. water, and the solution treated with 6 c.c. of hydrochloric acid of spec. grav. 1.124, or 9 c.c. of dilute (1 to 5) sulphuric acid, or enough to cause an acid reaction. Ten grams of pure dry beeswax are added, and melted, and the whole set aside to cool. The solidified mass is now carefully removed from the solution, dried with blotting-paper, and weighed; the weight being diminished by 10 grams gives the amount of fat acids [and resin]. 80 parts of fat acid indicate about 100 parts of good dried (soda) soap (HAGER); 11 parts fat acid represent an average of 12 parts of solid fat.—According to JEAN (*Chem. News*, xxvi., 206) the fat acids are estimated from the combined alkali (*i*), 12.6 parts of which (soda) unite with 100 parts anhydrous fat acids.

VOHL (*Jour. Chem. Soc.*, 1872, 934) separates the fat acids (and resin) by a limited quantity of *petroleum naphtha*. Ten grams of soap are dissolved in warm water, then decomposed by hydrochloric acid in a cylindrical vessel, and the solution, at 20° C., extracted with about 10 grams of petroleum naphtha. This solvent is afterward evaporated in a tared dish at 30° C., dried at 100° C., and the residue weighed as fat acids. As to Resins in this process, see *g*.

d. The fat acids may be approximately determined by the volume of their supernatant layer, after acidulation, in a graduated cylinder (BUCHNER). 1 c.c. equals 0.93 gram. The weight of fat acid plus $\frac{1}{16}$ equals the weight of fat; and 100 parts of fat correspond to 155 parts average hard soap.

e. PONS recommends a *volumetric determination of fat acid* by solution of calcium chloride; on which is based a valuation of the soap, taking average Marseilles soap—64 per cent. fat acids, 6 per cent. soda, and 30 per cent. water—as a standard or unit of value. One gram of this standard soap will precipitate 0.1074 gram calcium chloride (or 0.2532 gram barium nitrate); or 10. of soap, 1.074 of calcium chloride, the latter quantity being dissolved to make 1,000 c.c. [1.074 of calcium chloride may be obtained by dissolving 0.9675 of pure calcium carbonate, the solution being obtained exactly neutral.] Ten grams of the soap (carefully averaged) are dissolved in 100 c.c. of 85 per cent. alcohol—insoluble material being removed by decantation or filtration, and washing—and distilled water is added to make the liquid measure 1,000 c.c. In a stoppered flask of 60 to 80 c.c. contents, place 10 c.c. of the standard calcium solution, and add, from a burette measuring tenths c.c., the prepared soap solution, shaking after each addition until a foam remains on the surface (as in the soap test of hard waters). The number c.c. used contains as much soap as 10 c.c. of corresponding solution of standard soap would contain. Hence divide 10 by the number c.c. used, and the quotient expresses the value of the soap tested, as compared with the standard.

For the separation of the fat acids from each other, a work of difficulty, see under Fat Acids (55–57).

f. Uncombined fat can be extracted from soap (previously dissolved as far as possible in water) by petroleum naphtha at the temperature of 20° C. (Compare, under Butter, 63, *d* and *e*.)

g. Resin can be extracted from dried and pulverized soap by means of benzole, which dissolves only traces of the soap.—Or, the solution of fat acids in a little petroleum naphtha (a quantity equal to that of the soap)—as obtained by Vohl's process, given in *c*—contains the resin, which is now precipitated on diluting largely with petroleum naphtha. The precipitate subsides.—Also, when the fat acid (and resin) are treated with a mixture of water and a nearly equal volume of alcohol, the resin is dissolved out.

h. Soap may be precipitated from cold water solution by saturated solution of *common salt* (free from earthy bases), and washed on a filter with the same salt solution, with but little loss, the uncombined alkali (and alkaline carbonate) and the glycerine being contained in the filtrate and washings. The total of alkali in this filtrate may now be determined by volumetric solution of acid, showing *the uncombined alkali of the soap, including alkaline carbonate.*—If the soap is dissolved in alcohol, *alkaline carbonates* remain undissolved and may be determined by adding volumetric solution of acid to the residue.—*Free alkali* may be precipitated from alcoholic solution of soap by passing through a stream of carbonic acid gas.—A *qualitative test for free alkali* or alkaline carbonate is made by adding mercuric chloride to the soap solution; a red-brown to red-yellow precipitate indicates free alkali—the fat acid salts forming only white precipitates.

i. Then, for *volumetric determination of the combined alkali of the soap*, the soap precipitate is rinsed (with dilute solution of common salt) from the filter into a beaker, and decomposed by a five-times stronger than normal standard solution of (hydrochloric) acid, added to beginning of acid reaction. After which the fat acid may be separated as a cake and weighed, according to *c*—a weighed quantity of beeswax or paraffin being added, if necessary to secure solidification.

j. Determination of glycerin. Take 10 grams of soap, dissolve in alcohol, add alcoholic solution of sulphuric acid until precipitation ceases, and filter. Add baric carbonate and filter again. Evaporate until all the alcohol is expelled, and weigh the sweet residue as glycerin (SENIER). Or, treat the filtrate from acid precipitation of the fat acids with basic subacetate of lead, filter, remove the excess of lead by hydrosulphuric acid and filtration, neutralize with hydrochloric acid and extract with a mixture of alcohol 2 vols. and ether 1 vol. Evaporate this solvent and weigh as glycerin (VOHL).

k. A plan for determination of the constituents of soap, viz.: (1) Carbonates and other salts, color substances and foreign matters; (2) Free Alkali; (3) Combined Alkali; (4) Fatty acids with resin; (5) Fatty acids without resin; (6) Glycerin; (7) Water.*

For (1): Digest ten grams soap with alcohol (five or six ounces) on water-bath, filter and wash with hot alcohol in a hot funnel. Dry the residue at 100° C. and weigh. Analyze this residue by solution with water, by alkalimetry, etc.

For (2): Through the filtrate of (1) pass a stream of carbonic acid gas; if a precipitate forms, continue until its forma-

* SENIER: "A Process," etc., *Am. Jour. Phar.*, 1874, 353.

tion ceases; filter and wash and determine the alkali in the precipitate by a volumetric solution of (oxalic) acid. (See *h.*)

For (3): The filtrate from (2)—or if there was no precipitate in (2), the filtrate from (1)—after the addition of about an ounce of water, is evaporated on the water-bath to expel all the alcohol, and the (combined) alkali therein determined (as Soda or Potassa) by adding a normal solution of oxalic acid to acid reaction. (Compare *i.*)

For (4): To the mixture left in (3) add a little sulphuric acid; then add ten grams of previously melted beeswax, heat on a water-bath to fuse the wax, cool, weigh the cake, and subtract the weight of the wax. (Compare *c.*)

For (5): Dissolve 40 grams of soap in water, decompose by dilute sulphuric acid, cool at temperature below 14° C., separate and weigh the fatty acids; then digest them for some time with a mixture of water with nearly as much alcohol, until the subsident liquid (when the mixture has cooled and the fatty acids again solidified) ceases to be milky. Weigh the fatty stratum again; subtract the previous weight, and divide by four—for the resin in 10 grams soap. (Compare *g.*)

For (6): Proceed according to the first method under *j.*

For (7): Estimate by difference; or by evaporation of another portion with alcohol and sand, as directed in *a.*

68. RESINS. Compounds of **C**, **H**, and **O**. Vitreous and mostly brittle solids (when unmixed), softening and melting when gently heated, but not vaporizable (distinction from camphors); mostly heavier than water. The class includes some substances of pungent taste, some of poisonous effect, and some of intense color. Mostly insoluble or but slightly soluble in water: mostly soluble in absolute alcohol; by far the greater number soluble in ether and in benzole (means of separation from gums). Many resins are soluble in aqueous alkalies, by combination as resin-soaps; and in alcoholic solution show the acid reaction.

The *resins of commerce* include, first, vegetable exudates, of which the Resins proper mostly contain some extractive matters; the Gum-resins being mixtures with gums; the Oleo-resins, mixtures with volatile oils (including the source of common resin or colophony); and the Balsams, mixtures with volatile oils and acids formed by oxidation of volatile oils. Second, resins extracted from plants by alcohol, including some of both the Medicinal resins and the Color resins. And, third, resins obtained from liquid plant juices which are dried as a part of the manufacture; these including two bodies insoluble in alcohol, Caoutchouc and Indigo.

69. The *separation* of resins from volatile oils is effected by *distillation* with water; from gums, by *fusion* and straining at 100° C.; and from various bodies and from each other by action of the *solvents* applicable in the case. See Recapitulation, 99. Solution with alcohol and precipitation by pouring the solution into water is by far the most generally applicable process; solution with aqueous alkali and precipitation by acid may sometimes be employed.

70. The resinous matter of Aloes is fusible on the water-bath; insoluble in cold water, partly soluble in boiling water, freely soluble in alcohol, partly soluble in ether, scarcely at all soluble in chloroform, benzole, naphtha, bisulphide of carbon, freely soluble in aqueous alkalies and in glycerin.—Aloes yields *paracumaric acid*, as follows: The hot ammoniacal water solution is precipitated with acetate of lead, the filtrate freed from lead by dilute sulphuric acid, and this second filtrate is boiled in presence of the (excess of) sulphuric acid—forming (from resin) paracumaric acid in solution. The latter colors **ferric chloride** dark gold-brown.—The residue from an ammoniacal solution of material containing aloes, when saturated with hydrochloric acid, yields the odor of aloes. Farther, see Aloin.

71. Amber Resin. Amber contains Succinic acid, Volatile oil, and resin (two resins). Amber is a hard and brittle, more or less transparent solid, of spec. grav. 1.065; tasteless, aromatic when rubbed or warmed, of various colors, chiefly yellow or orange.—Subjected to gradually increasing heat, it softens; at 110° to 260° C., evolves a volatile oil colored blue by hydrochloric acid; at about 235° C., evolves succinic anhydride; at 287°, it fuses; at higher temperatures, yields first a colorless oil, then a yellowish wax.—Amber resin is insoluble in water, alcohol (except $\frac{1}{10}$ which is soft resin), ether, benzole, bisulphide of carbon, petroleum naphtha, volatile and fixed oils, but soluble in fixed alkalies (except a slight residue) and in concentrated sulphuric acid (with a red color).—Fuming **nitric acid** changes it to a nitrogenous resin of musk-like odor and gelatinous consistence—"artificial musk."

72. Ammoniac Resin. Ammoniac contains 72 per cent. resin and 22 per cent. gums, and a little volatile oil. Ammoniac is a solid, soft when warmed, brittle when cold, of specific gravity 1.207, whitish to yellow-brown and dirty gray, of a sweetish-bitter and acrid taste and strong peculiar odor. Ammoniac is partly soluble in water, alcohol, ether, acetic acid, and aqueous alkalies. Ammoniac Resin is wholly soluble in alcohol, in fixed and volatile oils, in sulphuric acid, acetic acid, and aqueous alkalies, and partly soluble in ether.

73. Assafetida Resin. Assafetida contains over 60 per cent of resin, about 30 per cent. of gums, and about 4 per cent. of volatile oil (whereon its odor depends). Assafetida is a solid, soft when warm, and brittle when cold, of spec. grav. 1.327, having an intense fetid and alliaceous odor and a bitter, acrid, and persistent taste. Its color is variegated and altered, being on fresh surfaces whitish to yellowish, becoming reddish to yellow-brown on exposure.—The volatile oil is separated by distillation with water, contains sulphur, and boils at 140° C.—Assafetida resin is readily soluble in alcohol, not wholly insoluble in water, nearly all soluble in ether, mostly soluble in alkalies.

74. Benzoin Resins. Benzoin or "benzoin-gum" consists of about three-fourths part resins, 10 to 15 per cent. of Benzoic acid, with a little gum and a very little volatile oil. Benzoin is a brittle solid, of spec. grav. above 1.062, melting and evolving benzoic acid when heated; of variegated colors, fragrant balsamic odor, and little taste, with slight acrid after-taste when chewed. Benzoin resins (three have been identified) are all soluble in alcohol, in concentrated **sulphuric acid** (from which **water** precipitates them violet), and in strong potassa solution, but insoluble in water. Resin-*a* is insoluble in aqueous carbonate of sodium, or in ammonia, but soluble in ether. Resin-*b* has the solubilities above given for *a*, except that it is insoluble in ether. Resin-*c* is sparingly soluble in ether and in volatile oils, and soluble in aqueous carbonate of sodium. The ether solution of *c* deposits a sediment which has been considered a fourth resin.—Dry distillation of benzoin, after removal of benzoic acid, gives a rose-red distillate.

75. Canauba Wax. Consists of myristic alcohol, resin, and other substance. It is a solid of spec. grav. 0.999, harder than beeswax, melting at 84° C., and of a greenish-yellow color. It is insoluble in water; dissolves with difficulty in alcohol, in ether, and in bisulphide of carbon; dissolves readily in oil of turpentine, but not at all in linseed oil, and not in aqueous alkalies. It is not changed by sulphuric acid, but is stained deep yellow by nitric acid.

76. Caoutchouc. Fusible at 120° C. (248° F.); not vaporizable. The larger part soluble in ether, benzole, bisulphide of carbon, petroleum naphtha, or oil of turpentine; wholly soluble in chloroform, and in a mixture of 100 parts bisulphide of carbon with 6 or 8 parts of absolute alcohol. Sparingly soluble in hot amylic alcohol. Not acted upon by alcohol or aqueous alkalies; slowly decomposed by concentrated sulphuric or nitric acid.

77. Colophony. Resin of Turpentine. Common Resin or Rosin.—A pale-yellow to brownish-yellow, translucent, brittle, vitreous solid, of spec. grav. of 1.07 to 1.08; softening at 70° C. and melting at 135° C. At a higher temperature it suffers destructive distillation, forming "essence of rosin" and then "rosin oil."—Insoluble in water; soluble in alcohol, ether, chloroform, benzole, petroleum naphtha (with much difficulty), volatile and fixed oils, methylic alcohol, aqueous alkalies (fixed and volatile), anilin, and hot aqueous carbonate of sodium. The three constituents—pinic, sylvic, and colopholic (or pimaric) acids—vary in solubility in certain solvents; cold dilute alcohol dissolving only pinic acid.

78. Copaiba Resins. Balsam of Copaiba consists of several resins and a volatile oil (a terpene). The most abundant of these resins, Copaivic acid (the proportion of which is very variable), is a brittle solid, crystallizable in colorless rhombs; soluble in strong alcohol, ether, benzole, petroleum naphtha, volatile and fixed oils, and aqueous alkalies. Its alcohol solution reddens litmus. Alcohol solutions of the alkaline copaivates, with alcohol solutions of salts of non-alkaline metals, on adding water, precipitate white metallic copaivates, more or less freely soluble in alcohol. The silver precipitate is crystalline, and the lead precipitate slightly so.—The *other resins* are soluble in alcohol, ether, fixed and volatile oils, and aqueous alkalies.

79. Copal. Spec. grav. 1.045 to 1.139. Brittle, softening at 50° C., more or less translucent, colorless to yellowish-brown. Consists of several resins. As a whole, it is imperfectly soluble in alcohol; slightly and slowly soluble in ether, bisulphide of carbon, ammonia; slowly soluble in oil of turpentine; readily soluble in oil of cajeput, or oil of rosemary, or "oil of caoutchouc." It is soluble in cold concentrated sulphuric and nitric acids, decomposing when these solutions are heated. Not soluble in alkalies; but combines with alkalies in boiling solution to form a soap soluble in water not containing free alkali.

80. Dammara Resin. *Australian.* Dammaric acid with Dammaran—that is, an acid and a neutral resin.—Both resins are soluble in absolute alcohol, ether, turpentine oil, benzole, petroleum naphtha, and solutions of fixed alkalies. The acid resin is soluble, the neutral resin insoluble in aqueous alcohol.—*East Indian* dammara (ordinary dammara). Spec. grav. 1.04

to 1.09, brittle, melting when heated. Partially soluble in absolute alcohol, about $\frac{3}{10}$ soluble in ether, fully soluble in fixed and volatile oils, benzole, and bisulphide of carbon, and in concentrated sulphuric acid with a red color. It is not soluble in aqueous alkalies.

81. Dragon's Blood. A brittle, dark-brown, opaque, odorless, and tasteless solid; soluble (with red color) in alcohol, ether, fixed and volatile oils, and mostly soluble in alkalies. The alcoholic solution forms red or violet precipitates with **metallic salts.**

82. Gamboge Resin. Gamboge is over three-fourths resin; the rest mostly gums, with a little starch. Gamboge is a brittle, pulverulent solid, of spec. grav. 1.22, burning when heated; reddish-yellow in mass, bright yellow in powder; nearly odorless at ordinary temperatures, but giving a peculiar odor when heated; a slight first-taste but a sweetish-acrid and dry after-taste when chewed, causing a flow of yellow-colored saliva.—Gamboge is easily emulsified with water, which dissolves gum from it, the resin being slowly deposited; is readily soluble in alcohol (with a little starchy residue), is soluble in aqueous alkalies, and yields its resin (only) to the solvent powers of ether, chloroform, bisulphide of carbon, and benzole (slowly). Boiling solution of sodic carbonate dissolves gamboge gelatinous. Gamboge is wholly dissolved by the successive action of ether and water (separation from commercial impurities).—*Gamboge Resin* ("gambogic acid"—usually extracted from gamboge by ether) is soluble in cold, concentrated, sulphuric acid, with a red color, and precipitated unchanged by adding water to this solution (a characteristic reaction). Boiled with **nitric acid** of 10 to 15 per cent. anhydride, the mixture then dissolved in alcohol and then treated with water, a yellow precipitate is obtained (distinction from Saffron or Turmeric).—The aqueous alkaline gambogates are precipitated red by **common salt**, and give red *precipitates* with baric salts, yellow precipitates with zincic and plumbic salts, brown precipitates with cupric salts, and brownish-yellow with argentic salts

—most of these precipitates being somewhat soluble in water and in alcohol.

For the *separation of gamboge resin* from associated medicinal resins (HAGER) the material is triturated with 98 per cent. alcohol (and pulverized heavy spar) at a gentle heat, and the extract so obtained is dried and digested with *chloroform*. Aloes resin, Convolvulin, and Colocynth resin are left behind (with a part of Agaric); while the gamboge resin is dissolved, with Jalapin, Guaiac resin, Myrrh, Tolu resin, Senna resin (and a part of Agaric). The residue from this chloroform solution is now digested with boiling solution of sodic carbonate; when, of those named above as in the chloroform solution, only the gamboge resin will dissolve (with traces of senna and agaric). Acids separate the gamboge resin from its soda solution.

83. GUAIACUM. A brittle, pulverizable solid, of spec. grav. about 1.2, melting at a moderate heat; of a faintly fragrant odor and persistent acrid after-taste. Its color is yellowish-green to reddish-brown; the former color induced by exposure to the air. Water dissolves a one-tenth of guaiac resin, strong alcohol about nine-tenths, alcohol of 83 per cent. slowly dissolves it all. Ether and oil of turpentine dissolve about as much as alcohol; benzole does not dissolve it. It nearly all dissolves in aqueous alkalies. **Sulphuric** acid dissolves it with a fine red color (and formation of glucose and guaiaretin); the solution is precipitated violet with water, or violet blue to blue-green by alcohol.—Guaiac resin is easy to suffer *oxidation*, whereby bright colors are produced. The powder and the alcoholic solution turn green by exposure to the air, or blue by exposure to ozone. The alcohol solution is also turned green by nitric acid, and blue by nitrous acid, chlorine, ferric chloride, or by ethereal solution of binoxide of hydrogen in presence of blood-stains. Hyposulphite of sodium changes the blue color to violet and then bleaches it; sulphurous acid bleaches it slowly—or promptly if zinc has been placed in the acid.

84. HEMP RESIN. Cannabin. Resin of Indian hemp.—A light-brown, lustrous solid or soft solid, melting at 68° C., and of a fragrant odor and bitterish, acrid taste. Insoluble in water, scarcely soluble in cold alcohol of 80 per cent., soluble in hot, strong alcohol, in ether, spirit of nitrous ether, chloroform, bisulphide of carbon, cold volatile oils, and warm fixed oils. Insoluble in aqueous alkalies; having a neutral reaction.

85. INDIGO BLUE. C_8H_5NO. Inodorous, tasteless, and neutral. Sublimes from the solid state, at about 288° C., without decomposition if in a current of air or in vacuum, forming purple-red vapors in open vessels, and condensing in right rhombic prisms.—It is insoluble in water, cold alcohol, ether, fixed and volatile oils when cold; hot alcohol and hot oil of turpentine and hot fixed oils dissolving it very sparingly. Insoluble in aqueous alkalies. Soluble in creosote and in hot phenic acid; soluble in concentrated sulphuric acid (as sulphindigotic acid).—Indigo blue is *separated* from fixed substances by sublimation from platinum foil (good indigo having 7 to 10 per cent. of ash); and by the use of solvents which leave it in residue. It is *valued*, in numerous processes, by the quantity of chlorine or other bleaching agent necessary to decolorize it.

86. JALAP RESINS. Resin of Jalap, of the pharmacopœias.—A brownish, brittle, opaque, fusible mass, or yellowish-gray to yellowish-white powder; of a repulsive odor, slight at ordinary temperatures, but much increased on heating, and a pungent, acrid taste.—It is soluble in alcohol (with neutral reaction), in aqueous fixed alkalies and alkaline carbonates, and in acetic acid; insoluble in volatile and fixed oils.—Resin of jalap consists of two distinct resins, Jalapin and Convolvulin; that of pharmacopœial or Tuberose jalap being about one-ninth jalapin and eight-ninths convolvulin; that of Fusiform jalap, mostly jalapin.

87. JALAPIN (or Scammonin) is a soft amorphous solid, brittle at 100° C., melting at 150° C., white in powder, tasteless, inodorous, and nearly neutral in reaction. It is very slightly soluble in water; freely soluble in ether, chloroform, methylic alcohol, benzole, petroleum naphtha, and oil of turpentine. Cold concentrated **sulphuric acid** dissolves jalapin; the solution becoming purple in five or ten minutes, then brown, and lastly black.—It dissolves in aqueous alkalies or their carbonates, and, on acidulating these solutions, Jalapic (Scammonic) acid is liberated—as a body soluble in water and having a strongly acid reaction. The salts of jalapic acid are nearly all soluble in water, but subacetate of lead precipitates it.—On heating Jalapin (or Jalapic acid) with dilute mineral acids, glucosic fermentation occurs, with formation of jalapinol and glucose. Jalapinol is insoluble in cold, sparingly soluble in hot water, soluble in alcohol and in ether; soluble in aqueous alkalies with combination as jalapinolic acid. Jalapinolic acid, liberated from its alkali salts by acidifying, is insoluble in water, but soluble in alcohol and in ether. Its lead and barium salts are nearly insoluble in water.—Jalapin and jalapic acid are amorphous; jalapinol *crystallizes* in white cauliflower-like masses, melting at 62° C.; jalapinolic acid crystallizes in tufts of needles (four-sided prisms), melting at 62° C.

88. Convolvulin (the larger portion of Tuberose jalap and a very small proportion of Fusiform jalap) is a brittle, vitreous solid, melting below 100° C. when moist, or at 150° C. when dry, colorless and transparent in mass, or white in powder, inodorous and tasteless, and of a slight acid reaction.—Nearly insoluble in water; soluble in alcohol, acetic acid, and aqueous alkalies and alkaline carbonates (as convolvulinic acid); not soluble in ether (separation from Jalapin).—It dissolves slowly in cold concentrated **sulphuric** acid, with a fine carmine-red color, afterward changing to brown; this change being a glucosic fermentation, with formation of convolvulinol and glucose. But dilute sulphuric acid has no effect.—Convolvulic acid is formed in acidifying the alkaline solutions of convolvulin; it is a white solid, fusing above 100° C., having a strong acid reaction, and freely soluble in water and alcohol, insoluble in ether. Its metallic salts are soluble, except that formed with basic acetate of lead.

89. Lac Resin. Stick Lac consists of about two-thirds resin, one-tenth coloring matter, with wax, gluten, etc. Seed

Lac contains more resin and less coloring and nitrogenous matter. Shell Lac is about 90 per cent. resins, 5 per cent. wax, 2.5 per cent. gluten, and 0.5 per cent. coloring. The coloring matter of lac is soluble in water; is bright red with acids and deep violet with alkalies; is precipitated by alum.

Shell Lac is insoluble in water; soluble in alcohol; mostly soluble in methylic alcohol; wholly soluble in aqueous alkalies, and in water solution of borax, and in hydrochloric and acetic acids.—Lac resin is *separated* from most other resins, and from many natural and commercial impurities, by dissolving in a solution of ½ part borax and 20 to 30 parts water to one part of lac. The solution may be diluted farther. (Good shell lac leaves not over 1.5 per cent. residue; poor, as much as 8 per cent.) By 10 per cent. ammonia at 25° to 30° C. lac is not dissolved, while Colophony dissolves and appears, after acidulation, as a precipitate. Cold ether (of 0.720 spec. grav.) does not dissolve more than 5 to 6 per cent., chloroform not over 7½ per cent. from good lac, the dissolved part being wax with a very little resin (separation from Colophony and other resins).

90. Mastic. A translucent solid, brittle and inodorous at ordinary temperatures, but soft and ductile when chewed and fragrant when heated, of a faintly terebinthinate taste. Alcohol dissolves about four-fifths, leaving Masticin undissolved. Ether, chloroform, and oil of turpentine dissolve it wholly. It is largely soluble in benzole.

91. Myrrh Resin. Consists of resins, about ⅓ part; gums, about ⅔ part; with a very little soluble extractive. Myrrh forms an emulsion and partial solution with water, a nearly complete solution with much aqueous potassa, and yields its resin to alcohol, ether, and chloroform.—The Resin of Myrrh is readily soluble in alcohol, ether, chloroform; slightly soluble in hot solution of sodic carbonate; about one-half part soluble in bisulphide of carbon. That part extracted with bisulphide of carbon, when dissolved in alcohol and warmed with 25 per cent. **nitric acid**, gives a violet color.

92. OLIBANUM Resin. Frankincense. Incense.—Olibanum is about one-half part resin, one-third part gum, one-twelfth part volatile oil. The gum is soluble in water; the resin is soluble in alcohol.

93. Resin of PERU Balsam. About $\frac{1}{4}$ resins, $\frac{2}{3}$ volatile oil, less than $\frac{1}{10}$ cinnamic acid. The Balsam is of thick-syrupy consistence; spec. grav. 1.15 (sinks in an 18 per cent. solution of common salt). Soluble in absolute alcohol in all proportions, or in 6 parts of 90 per cent. alcohol with slight turbidity; perfectly soluble in all proportions of absolute ether, chloroform, and amylic alcohol. Bisulphide of carbon dissolves the greater part; benzole and petroleum naphtha dissolve about one-half. It mixes with about $\frac{1}{8}$ part of castor oil, and with $\frac{1}{4}$ part copaiba balsam. Sulphuric acid converts the balsam into a thick red mass. Aqueous alkalies dissolve out the resin. 10.0 of the balsam requires over 0.7 grams crystallized sodic carbonate to neutrallize its cinnamic acid.

94. PODOPHILLUM RESIN. Consists of two resins. Insoluble in water; wholly soluble in alcohol; about $\frac{3}{4}$ part soluble in ether; wholly soluble in aqueous alkalies, from which solutions acids precipitate it (distinction from resins of Jalap and Scammony). Insoluble in benzole.

95. SANDARAC. A brittle, yellow solid. Contains three resins. Sandarac is insoluble in water; wholly soluble in alcohol—$\frac{3}{4}$ part dissolving easily in cold ordinary alcohol, a small part requiring boiling alcohol, and a still smaller part a large quantity of this solvent for solution. It is easily soluble in ether and in oil of turpentine, imperfectly soluble in bisulphide of carbon, benzole, petroleum naphtha, or linseed oil. Nitric acid colors it clear brown.

96. SCAMMONY Resin. Convolvulin. See Jalapin (87).

97. Resinous part of STORAX. Consists of (two) resins, and Styracin or Cinnamate of Cinnyl ($C_9H_9C_9H_7O_2$). Alcohol and ether dissolve the whole. In cold alcohol, the styracin crystallizes in tufts of prisms. Styracin is tasteless and odorless, more freely soluble in ether than in alcohol. Treated with hot nitric acid, or with chromic acid, or with sulphuric acid and binoxide of manganese, it yields benzoyl hydride (oil of bitter almonds).

98. Resins of Tolu Balsam. The Balsam consists of 80 to 90 per cent. of resin, about 12 per cent. of cinnamic acid, and less than 1 per cent. of volatile oil. It is wholly soluble in alcohol, chloroform, volatile oils, and aqueous alkalies; partly soluble in ether; insoluble in benzole, petroleum naphtha, bisulphide of carbon, and solution of carbonate of sodium. The Resins of Tolu balsam are soluble in cold concentrated sulphuric acid, without change.

99. Separation of Resins by Solvents. Recapitulation.—**Water** *dissolves* a part of the resin of Assafetida, a part of Gamboge, about $\frac{1}{10}$ of Guaiac resin, and slightly dissolves Jalapin.

a. **Alcohol** *fails* to dissolve $\frac{9}{10}$ of Amber, Canaüba wax, Caoutchouc, a part of Copal, $\frac{1}{10}$ of Guaiacum, Indigo blue (dissolving slightly with heat), and $\frac{1}{5}$ of Mastic.

b. **Aqueous Alkalies** (potassa or soda) *dissolve* Aloes resin, Amber, Ammoniac, Assafetida (mostly), Benzoin, Colophony, Convolvulin (with change), Dammara (Australian), Dragon's Blood (mostly), Guaiacum, Jalapin (with change), Lac resin, Myrrh, and resins of Podophyllum and of Peru and Tolu balsams.—These solvents do *not* dissolve Canaüba wax, Caoutchouc, Copal, Dammara (East Indian), Hemp resin, Indigo blue.

c. **Ether** *dissolves* resin of Aloes, Ammoniac (in part), Assafetida resin (mostly), Benzoin (in part), Canaüba wax (with difficulty), Caoutchouc (mostly), Colophony, Copal (with difficulty), Dammara (in part), Dragon's Blood, Gamboge, Guaiacum (in greater part), Hemp resin (Cannabin), Jalapin, Mastic, resin of Peru balsam, $\frac{3}{4}$ of Podophyllum resin, Sandarac, Styracin, and resin of Tolu balsam.—Ether does *not* dissolve Amber, Indigo, and $\frac{1}{4}$ of Podophyllum resin.

d. **Chloroform** *dissolves* Caoutchouc, Colophony, Gamboge, Guaiacum, Hemp resin (Cannabin), Jalapin, Mastic, Myrrh, resin of Peru balsam, resin of Senna, resin of Tolu balsam.—Chloroform does *not* dissolve Agaric (in chief part), resin of Aloes, resin of Colocynth, Convolvulin.

e. **Bisulphide of Carbon** *dissolves* Canaüba wax, Caoutchouc, Copal (slowly), Dammara, Gamboge, Hemp resin, ½ of Myrrh, resin of Peru balsam, Sandarac (in part).—It does *not* dissolve Amber, Indigo blue, ½ of Myrrh, resin of Tolu balsam.

f. **Benzole** *dissolves* Caoutchouc, Colophony, Dammara, Gamboge, Jalapin, Mastic (mostly), ½ of the resins of Peru balsam, Sandarac (in part). Benzole does *not* dissolve Amber, Guaiacum, resin of Podophyllum, resin of Tolu balsam.

g. **Oil of Turpentine** *dissolves* Ammoniac, Benzoin resin (in part), Canaüba wax, Caoutchouc, Colophony, Copal (slowly), Dammara, Dragon's Blood, Guaiacum (mostly), Hemp resin, Jalapin, Mastic, Sandarac, resin of Tolu balsam.—It does *not* dissolve Amber, Indigo (without heating).

h. **Sulphuric Acid,** concentrated, cold, *dissolves* Amber (with red color), Ammoniac, Benzoin resin, Convolvulin (with red color turning brown), Copal, Dammara (with red color), Gamboge (with red color), Guaiacum (with red color, etc.), Indigo blue.—It does not dissolve Caoutchouc.

100. VOLATILE OILS. In *composition,* 1st, Hydrocarbons, or "elæoptenes," mostly of the formula $(C_{10}H_{16})n$, a large class;

2d, Oxidized oils (C, H, O), including (1) hydrates of hydrocarbons, the "stearoptenes" or camphors, a moderate number being found alone and a large number in mixtures with the elæoptenes, (2) aldehydes, (3) compound ethers, generally in natural mixture with elæoptenes, (4) of irregular composition;

3d, Sulphurized oils (C, H, O, S), a small class, products of natural fermentation, and having odors resembling each other.

101. Mostly *liquids,* a few oils and stearoptene parts of oils melting at a little above ordinary temperature; the greater number lighter, a few heavier, than water; very slowly volatile at ordinary temperatures, mostly having boiling points above 150° C., but all distilling, slowly, with steam at 100° C., and leaving a transient oil-spot on paper. They are noted for strong and persistent odors; colorless, or with pale colors, in a few instances tinted blue with cœrulein, transparent and possessed of strong refractive powers.—The volatile oils are neutral in *reaction;* not generally liable to decomposition or combination except with oxygen. By air and light many of them alter and form resinous bodies; the elæoptenes forming stearoptenes, and (by oxidizing agents) aldehydes forming acids.

102. Volatile oils are very sparingly *soluble* in water, requiring intimate mixture and generally from 600 to 1,000 parts of water for solution; soluble in alcohol, and in all proportions of absolute alcohol, ether, chloroform, benzole, petroleum naphtha, bisulphide of carbon, fixed oils and other volatile oils. Alkalies do not affect them.—Certain oils, after distillation with water, retain traces of water in solution. This occurs with oils of bergamot, cinnamon, cloves, juniper, lavender, lemon, rosemary, sassafras, spike, wintergreen; not with oils of amber, cedar, rue, turpentine. The presence of water is shown by turbidity on mixture with several volumes of petroleum naphtha (Leuchs).—Volatile oils are scarcely at all soluble in aqueous solutions of chloride, nitrate or sulphate of sodium.

103. The volatile oils are *characterized* by their individual odors, their physical properties (as stated above and in 105 and 106), by various special reactions (the most of which are stated in 107 to 114), by their refractive indices and their absorption spectra, and by their cohesion-figures when dropped upon a still surface of pure water.*

104. Volatile Oils are separated from substances more or less volatile by their distillation with steam; from many substances by their slight solubility in **water** (farther lessened by

* Tomlinson, Moffat: *Chem. News*, 1869. Crane: *Am. Jour. Phar.*, 1874, Sept., and *Phar. Jour.*, 1874, p. 242, *et. seq.*

common salt) and ready solubility in alcohol, ether, etc.—*From Fixed Oils* they may be separated by distillation with water; by solution in alcohol (not from castor oil); or by alkaline **saponification** of the fixed oil.

From Alcohol, they may be separated (in greater part) by addition of **water**; (in part) by addition of **fixed oil**; (in part) by addition of dry **chloride of calcium**, and (with a little loss) by repeated **distillations** with water.—Also (qualitatively) by adding to 5 or 10 drops of the oil, in a test-tube, a fragment of dry **tannic acid**, agitating, and leaving several hours at ordinary temperature. In absence of alcohol, the tannin remains solid, porous, and floats; in presence of alcohol, it becomes pasty or liquid, and adheres to the glass or sinks (HAGER).—Farther, volatile oils may be (quantitatively) separated from alcohol by **glycerin** (HAGER): In a graduated cylinder place 10 parts of the mixture of oil and alcohol and 10 parts of a mixture of $\frac{2}{3}$ glycerin and $\frac{1}{3}$ water, agitate, and set aside a few hours for separation. Read off at about 17.5° C. (Oil of Balm is soluble in glycerin.)—Separation of volatile oils (or of Camphor) from alcohol may be made by water solution of **nitrate or sulphate of sodium** much more nearly than by water alone, and for approximately quantitative purposes. In a flask with a graduated neck, or a wide cylinder having its upper third narrowed and graduated, place about 3 vols. of a half-saturated solution of the salt and add 1 vol. of the alcohol solution of oil or camphor, agitate thoroughly, add enough of the salt solution to adjust the surface to graduated portion of the measure, and set aside at 20° to 25° C. until the liquids separate clear. The c.c. of oil multiplied by its spec. grav. equal the grams. For camphor (and if desired for oils) the process may be completed gravimetrically, by adding about 3 parts of exactly weighed paraffin, fusing (inserting a platinum hook), and weighing when cold. Compare 67, *c*.

105. Color and Specific Gravity of Volatile Oils.

Volatile Oils.	Color of the Crude Oil.	Color after Rectification.	Spec. Grav.
Amber, . .	Yellowish or reddish-brown.	Colorless or yellowish.	0.80—0.88
Anise, . . .	Pale yellow to yellow.		.98— .99
Balm, . . .	Yellowish.		.85— .89
Bergamot . .	Yellowish-green or brown-yellow.	Colorless or yellowish.	.88— .95
Bitter Almond, .	Yellowish, growing darker.		1.04—1.06
Cajeput, . .	Green.	Colorless.	.91— .94
Calamus, . .	Pale yellow.		.89— .95
Camphor (oil of).	Yellowish to reddish-brown.		.94
Caraway, . .	Pale yel'w, growing brownish.	Colorless.	.91— .94
Cardamom, .	Greenish-yellow.		.93— .95
Cascarilla, . .	Dark yellow.		.90— .93
Chamomile, .	Dark blue.		.01— .04
" Roman, .	Light blue.		
Cinnamon, . .	Yellow, becoming darker.		1.03—1.06
" (Cassia), .	Light yellow to dark yellow.		1.02—1.08
Cloves, . .	Brownish-yellow.		1.03—1.06
Copaiba, . .	Colorless or yellowish.	Colorless.	.87— .91
Coriander, . .	Yellowish.	Colorless.	.87— .89
Cubeb, . .	Colorless.		.92— .94
Cummin, . .	Yellowish.		.90— .97
Dill, . . .	Yellowish, becoming red-br'n.		.88— .93
Eucalyptus, .	Colorless.		.88— .93
Fennel, . .	Colorless, growing yellowish.		.90— .99
Galbanum, . .	Yellowish.		.90— .92
Galangal, . .	Yellowish.		.91— .92
Geranium, . .	Yellowish.		.90— .91
Hedeoma, . .	Light yellow.		.94
Hops, . . .	Pale brownish yellow.		.90— .91
Jasmin, . .	Yellowish.		
Juniper wood, .	Colorless or yellowish-green.		
" berries, .	Colorless, yel'wish or greenish.	Colorless.	.84— .89
Lavender, . .	Colorless, growing darker.		.87— .90
Lemon, . .	Yellowish.	Colorless.	.845-.865
Mace, . . .	Pale yellow.	Colorless.	.87— .95
Marjoram, . .	Clear yellow.		.89— .92
Myrrh, . .	Colorless or yellowish.		1.10—1.13
Nutmeg, . .	Pale yellow, darkening.		.90— .93
Orange flowers, .	Colorless, growing yellowish.		.85— .90
Orange peel, .	Yellowish.	Colorless.	.83— .85
Origanum, . .	Yellowish to brown-yellow.		.80— .90
Parsley, . .	Yellowish.		1.02—1.04
Pepper (black), .	Yellowish to clear-brown.		.85— .89
Peppermint, .	Pale yellow, or greenish iridescent.		
Pimento (allspice)	Colorless to yellowish.		.89— .92
Rosemary, . .	Colorless or pale yellow-green.		.88— .92
Roses, . . .	Colorless, reddish, or yel'wish; concrete below 20° C.		.83— .84
Rosewood, . .	Pale yellow.		

105. Color and Specific Gravity of Volatile Oils.—*Continued.*

Volatile Oils.	Color of the Crude Oil.	Color after Rectification.	Spec. Grav.
Rue, . . .	Yellowish.		.85— .90
Sage, . . .	Green-yellow or yellowish.		.86— .92
Sassafras, .	Yellowish to red-yellow.		1.06—1.08
Savine, .	Colorless or yellowish.		.89— .93
Spearmint, .	Yellowish, becoming dark, red-brown.		.91— .98
Tansy, .	Pale yellow or green yellow.		.90— .95
Turpentine,	Colorless.		.87— .89
Thyme, .	Yellow-green, red-brown.		.87— .89
Valerian, .	Yellow-brown, green-brown.		.90— .96
Wintergreen, .	Reddish.	Colorless.	1.14—1.17
Wormseed (Santonica), .	Brownish-yellow.		.91— .96
Wormwood,	Green.		.88— .93
Yarrow, .	Dark-blue.		.87— .92
Ylang-Ylang,			.98

106. Solubility of Volatile Oils in Alcohol of sp. gr. 0.822 (90 per cent.)

Take, in a test-tube, from a minim measure, 5 or 10 minims of the oil, and then as many minims of the alcohol as required, with agitation, to dissolve. The oils which form solutions more or less turbid are given with figures in heavy type. It will be borne in mind that oils are less soluble when old than when fresh. Also, that mixtures of oils usually have solubilities midway between those of the individual oils therein.

Alcohol required, at 17° to 20° C., for 1 vol. of oil of

Oil	Alcohol	Oil	Alcohol
Amber, ...	3½ vols.	Cinnamon, ...	1 vols.
Anise, ...	1 "	" (Cassia), . .	1 "
Balm, ...	3 "	Cloves, ...	1 "
Bergamot, ...	½ "	Copaiba, ...	0 "
Bitter Almond, ...	1 "	Cubeb, ...	**25** "
Cajeput, ...	11 "	Cummin, ...	1 "
Calamus, ...	1 "	Fennel, . . .	1 to 2 "
Caraway, . . .	½ to 1 "	Juniper berries, . .	**10** "
Cardamom, . . .	½ to 1 "	Lavender, ...	1 "
Chamomile, ...	8 "	Lemon, ...	**50** "

Mace, ...	5 vols.	Rue, ...	1 vols.
Marjoram, ...	1 "	Sage, ...	1 "
Orange flowers, . . .	1 to 2 "	Savine, ...	1 to 2 "
Orange peel, ...	5 "	Tansy, ...	1 "
Parsley, ...	3½ "	Turpentine, ...	9 "
Peppermint, ...	1 "	" rectified, .	10 to 12 "
Rosemary, ...	1 to 2 "	Valerian, ...	1 "
Roses, ...	50 to 70 "	Wormwood, ...	1 "

107. Reaction of Volatile Oils with Iodine and Bromine. (1) When about 0.1 gram of dry pulverized **iodine** is placed at ordinary temperature in a watch-glass and 4 or 5 drops of the oil are dropped upon it:

(a) *Giving instantaneous reaction, with much heat and strong effervescence;* Oils of

Bergamot,
Eucalyptus,
Hops,
Lavender,
Lemon,
Mace,
Orange flowers,
Orange peel,
Savine,
Turpentine,
Wormwood (old).

(b) *Generating slight heat, with gentle effervescence:* Oils of

Anise,
Balm,
Caraway,
Chamomile,
Cubeb,
Dill,
Fennel,
Juniper,
Marjoram,
Rosemary,
Sage,
Sassafras,
Thyme.

(c) *Giving no reaction, or very slight:* Oils of

Amber,
Bitter Almond,
Cajeput,
Calamus,
Cascarilla,
Cinnamon (Ceylon),
Cinnamon (Cassia),
Cloves,
Mustard,
Parsley,
Peppermint,
Roses,
Rue,
Sassafras,
Thyme,
Valerian,
Wormwood (fresh).

(2) Upon 5 or 6 drops of the oil, on a watch-glass, one drop of **bromine** is let fall (Maisch).

(*a*) Giving *detonation* with Oils of

Amber,
Bergamot,
Hedeoma,
Juniper berries,
Juniper wood,
Lemon,
Turpentine.

(*b*) Giving a *hissing sound* with Oils of

Anise,	Sassafras,
Caraway,	Wormseed.

(3) To 5 or 6 drops of the oil, on a watch-glass, add 5 drops of **ether solution of bromine** (1 vol. bromine to 5 vols. officinal ether, added slowly, while cooling, just before use).*

(a) *Vapors evolved* with Oils of

Copaiba (green color ; afterward brownish-green with brown sediment).

Cubeb (violet color, deepening ; afterward dark greenish-blue, with violet-black sediment).

Orange peel (yellow color soon appears ; afterward pale brown and transparent).

Patchouli (deep violet color, deepening ; sediment dark brown).

Sassafras (at first cloudy ; afterward pale brownish-yellow).

Spearmint, old, yellowish-red (color changes to yellowish-brown ; sediment lighter).

Wintergreen (formation of a resinous white substance, spreading over the glass).

(b) *Vapors not evolved* with Oils of

Anise (white color ; with more bromine, yellowish-red).

Bergamot (color greenish-brown yellow, then reddish-brown yellow).

Bitter Almond (dissolves without reaction ; after evaporation of the ether, two liquids separate—one deep, the other light red).

Cajeput (supernatant liquid scarcely colored ; appearance of green droplets).

Calamus (colors red-brown, brown-green ; finally a dark sediment).

Caraway (little reaction ; sediment yellowish-brown).

Cinnamon (color lemon-yellow, turning to amber-brown).

Cloves (color greenish ; lower stratum alters to pale grayish-black).

Hedeoma (color changed to purplish and darkened ; liquids not miscible).

Lavender (light greenish, darkening to deep sea-green).

Lemon, old (brisk reaction ; colors reddish-yellow and greenish).

Mustard (miscible, colorless ; afterward milk-white).

* MAISCH : *Proc. Am. Phar. A.*, 1859, 338.

Nutmeg (at first colorless ; the lower stratum then brownish and milky to clear).

Peppermint (colors—yellowish, then reddish, then brown—thickening).

Rosemary (colorless ; afterward lower stratum is light-brown).

Rue (at first cloudy, then pale brownish yellow).

Valerian (at first purplish-black ; then upper stratum deep violet, lower greenish-black, marginal blue and red spots).

Wormseed (reaction is slow ; heavier liquid red to brown ; lighter liquid light brown and almost clear).

Wormwood (darkens a little without movement).

108. Reaction of Volatile Oils with Sulphuric Acid and Alcohol (HAGER's Method). In a test-tube of about 1.3 centim. (0.5 inch) diameter, 5 or 6 drops of the oil are agitated with 25 to 30 drops of concentrated **sulphuric acid,** after which it is noted how much heat and how much turbidity, if any, have been produced. When the liquid, if heated, has cooled again, 8 or 10 c.c. of 90 per cent. **alcohol** are added, with brisk shaking while the test-tube is closed by the finger. Now the production of color and of turbidity are noted. In case of turbidity, after standing, a subsident layer usually appears, having a characteristic color, and being soluble in cold or in hot alcohol or in chloroform.

(*a*) The mixture of oil with acid and alcohol, is *clear and transparent, or but very slightly turbid,* in case of Oils of

Amber (with sulphuric acid, *not heated,* dark yellow and turbid ; after adding alcohol, yellow, slightly turbid, made clear by boiling).

Anise (with the acid, in part dark red and thick, and in part clear and limpid ; with the alcohol the thick part remains dark and undissolved, while the liquid part is clear and nearly colorless).

Bitter Almond (with the acid, a brown color and much heat without turbidity ; with the alcohol, a clear and nearly colorless mixture).

Cloves (with the alcohol, the mixture is nearly or quite clear).

Dill (with acid, generation of heat and vapors, with dark yellow-red color and some turbidity ; with alcohol, a pale cinnamon-brown mixture, nearly or quite clear—fully clear on boiling).

Fennel (with acid, heat and vapors, the mixture dark red and pretty clear ; with alcohol, yellowish, clear solution).

Mustard (with acid, *very little heat*, yellowish tint, clear ; with alcohol, colorless and clear).

Nitrobenzole or "artificial oil of bitter almonds" (without turbidity).

Peppermint, best (with the acid, *slight heat* and yellow-red color ; with the alcohol, light red, slightly turbid mixture, made clear by boiling).

Peppermint, American (with the acid, heat and dark brown-red color ; with the alcohol, brownish and turbid, made clear by boiling).

Roses (with acid, heat, thick vapors, and dark brown-red color ; with alcohol, brown, clear, and transparent).

Valerian (with the acid, heat and slight vaporization, dark red color, slight turbidity ; with the alcohol, red, turbid, but rendered clear by boiling).

(*b*) The mixture of oil with acid and alcohol is left *more or less turbid*, in case of Oils of

Balm (with acid, heat, vapors, brown-red color, and turbidity ; with alcohol, cinnamon-brown, somewhat turbid ; after boiling becomes clear with separation of dark drops).

Bergamot (with acid, heat and vapors ; the alcohol solution pale grayish-yellow turbid, with flocculent separate after shaking; after one or two days, the residue is but slight and divisible on shaking, the liquid being clear yellow).

Cajeput (with acid, heat and vapors, light yellow color and turbidity ; with alcohol, pale rose-gray turbidity, made clearer by boiling).

Caraway (with acid, heat and vapors, dark yellow to red-brown color, turbidity ; with the alcohol, a red and turbid mixture, made nearly clear by boiling).

Cascarilla (with acid, heat and vapors, dark brown-red color, turbidity ; with alcohol, the same ; an hour after boiling, dark brown-violet to bluish-red).

Cinnamon (Cassia) (with acid, a strong heat and vaporization, dark black-brown, very thick mixture ; after the alcohol, the dark viscid mass remains mostly insoluble, with a milky olive-green liquid above).

Copaiba (with the acid, heat and vapors, the color dark yellow-red, with turbidity ; with alcohol, red and turbid, not made clear by boiling).

Coriander (with sulphuric acid, heat and vapors, dark red color, scarcely turbid ; with alcohol, dark brown, with green shade, and turbid).

Eucalyptus (with sulphuric acid, heat and vapors, light reddish-yellow color, with turbidity ; with alcohol, very turbid, with whitish-peach-blow or pale rose-gray color).

Geranium (with acid, much heat and thick vapors, turbid, dark yellow-red; with alcohol, turbid and dark brown ; after boiling, turbid and red-brown).

Juniper berries (with acid, heat and vapors, turbid, dark-yellow-red ; after the alcohol, very turbid, sometimes flocculent, of blackish-rose color ; after boiling, turbid ; after a few hours, a light-colored resinous mass separates).

Juniper wood (with acid, heat and vapors, turbid, orange-red ; with alcohol, pale yellowish, turbid before and after boiling).

Lavender (with acid, heat and vapors, turbid and brown-red ; with alcohol, turbid, dark brown with green tint).

Lemon (like Bergamot oil : after one or two days, the slight residue forms opaque yellow drops not divisible by shaking).

Mace (with acid, heat and vapors, turbid, dark red ; with alcohol, turbid and dark reddish-brown, not made clear by boiling).

Marjoram (with acid, heat without vapors, turbid and yellow-red ; with alcohol, very turbid, peach-blow and almost milky ; turbid after boiling).

Orange flowers (with acid, heat and vapors ; after alcohol, turbid and brown, approaching red ; after boiling, a little darker and less turbid).

Orange peel (with acid, a strong heat, turbidity and red-brown color ; with alcohol, whitish-yellow; turbid before and after boiling).

Parsley (with acid, a moderate heat and a little vapor, very dark red ; with alcohol, very turbid, red, with swimming flocks).

Rosemary (with acid, strong heat but no vapors, yellow-red and turbid ; with alcohol, milky turbid ; turbid after boiling).

Rue (with acid, heat and vapors, dark red, turbid ; with alcohol, raspberry-red, turbid ; clear after boiling).

Sage (like Oil of Rue).

Savine (with acid, strong heat without vapors, moderately turbid, dark red; with alcohol, turbid, reddish-clay-colored ; after boiling, less turbid, pale red).

Tansy (with acid, heat and vapors, dark red, turbid with alcohol, yellow-red, less turbid ; after boiling, clear).

Thyme (with acid, heat and vapors, red, turbid after alcohol ; after boiling, clear, with swimming oil-drops).

Turpentine (deviating greatly from differences of production and of age).

Wormseed (Santonica) (with acid, moderate heat and vapors, dark red, turbid ; with alcohol, cinnamon-brown, turbid ; becoming clear on boiling).

Wormwood (with acid, heat and vapors, red-brown, turbid ; with alcohol, dark, green-violet, opaque, turbid ; becoming clear with more alcohol).

Ylang-Ylang (with acid, heat and vapors, turbid and dark red ; with alcohol, pale brick-red and very turbid, less turbid after boiling).

109. Reaction of Volatile Oils on Sulphide-of-Lead-Paper (G. Williams). Blotting-paper is wetted in a dilute alcoholic solution of **acetate of lead** and dried in an atmosphere of **hydrosulphuric acid.** A few drops of the oil are let fall on a strip of this paper, which is placed in a (dry) dark place for 5 or 10 or 15 hours, when the degree of bleaching is noted.

The paper *is bleached* by Oils of Lavender, Peppermint, Rosemary, Turpentine. The paper is *not bleached* by Oils of Anise, Bergamot, Cajeput, Cinnamon, Juniper berries, Lemon, Orange peel, Sage, Thyme.

110. Reaction of Volatile Oils with Sodium (Dragendorff). The Hydrocarbons are not affected; the Oxidized oils are more or less readily decomposed. Ten drops of the oil are treated with a small piece of the metal. The result is discovered after 5 or 10 minutes. (Alcohol causes a prompt reaction, with evolution of hydrogen.) Little or no change occurs with Oils of Amber, Bergamot, Copaiba, Lavender, Lemon, Nutmeg, Pepper, Peppermint, Rosemary, Sage, Turpentine. Oil of Mustard evolves hydrogen.

111. Identification of Resinified or Old Oils, or of Resins or Fixed Oils in mixture with volatile oils. Evaporate 1 gram of the oil, on a tared watch-glass, at 70° to 90° C. (or over the water-bath). Fresh and unchanged oils, free from mixture, leave only a scarcely perceptible and not weighable residue.

This residue, fully freed from volatile oil, may be tested for *Castor Oil*, by treatment for œnanthyc acid, as described under Ricinoleic Acid (46).

112. Identification of Turpentine Oil. The sparing solubility of this oil in aqueous alcohol affects its mixtures with other oils, but does not enable it to be separated. The alcohol should be 75 to 90 per cent.—HEPPE's test is with **nitroferricyanide of copper**—prepared by precipitating solution of sulphate of copper with solution of nitroferricyanide of sodium, and washing and drying the precipitate. In a test-tube place a bit of this reagent as large as a pea, then about 25 drops of the oil, and heat, so as finally to boil for a few seconds, and set aside to subside. Turpentine oil (also lemon oil) does not suffer change, or more than slight change—while the sediment of nitroferricyanide is green or blue-green. Other volatile oils are darkened to different colors; while the sediment of copper salt is gray, brown, or black.

113. Identification of Valerian Oil. One drop of the oil is dissolved in 15 drops of bisulphide of carbon, then shaken with sulphuric acid, and afterward one drop of nitric acid, of spec. grav. 1.2, is added. A fine blue color results when even slight portions of the oil are present (FLUCKIGER).

114. Identification of Oil of Peppermint. 50 to 70 drops of the oil, with 1 drop of nitric acid, of spec. grav. 1.20, turns faintly brownish, and after an hour or two becomes fluorescent—blue-violet or green-blue by transmitted and copper-color by reflected light (FLUCKIGER).—Chloral hydrate, on contact with oil of peppermint, colors it reddish. The tint deepens to cherry-red, is intensified by sulphuric acid, and varied to dark violet by chloroform. (No color is obtained with oils of lemon, bergamot, juniper, rosemary, cloves, anise, or fennel.)

For qualitative separation of Benzole from volatile oils, see 119; of Nitrobenzole from Bitter Almond Oil, see 120.

115. CAMPHOR. $C_{10}H_{16}O$. Laural Camphor.—A slightly unctuous, pellucid solid, friable with cleavage, of specific gravity 0.985 to 0.996; melting at 142° C. (288° F.), slowly vaporizable at ordinary temperatures, condensing in hexagonal plates, boiling at 204° C. (400° F.) It is *soluble* in 1,000 parts of water applied by ordinary contact, or in 150 to 200 parts of water by trituration with an insoluble powder; freely soluble in alcohol, ether, chloroform, benzole, petroleum naphtha, methylic alcohol, amylic alcohol, creosote, acetic acid, mineral acids, bisulphide of carbon, fixed and volatile oils, and forms a liquid mixture with solid chloral hydrate.—Minute particles of camphor, dropped upon water, rotate, with velocity in proportion to their smallness. If an oiled pin-point is then touched to the water, the rotations are stopped, and the camphor particles carried out by the enlarging circular oil-film.

By prolonged boiling with concentrated **nitric acid** or permanganate of potassium, camphor is changed into Camphoric Acid. The latter is sparingly soluble in water, from which it crystallizes in colorless scales or needles, of sour and bitter taste, melting at 70° C., and forming insoluble salts with lead and many other metals.—By heating in a closed vessel with **bromine**, Bromated Camphor is formed, as a crystallizable solid, not soluble in water.

116. CREOSOTE. Chiefly Creosol, $C_8H_{10}O_2$, and Guaiacol, $C_7H_8O_2$. An oily limpid *liquid*, of spec. grav. 1.060 to 1.085, colorless or yellowish (growing brownish in the light), boiling at 200° to 206° C. (392° to 403° F.), having a neutral reaction, a strong and persistent smoky odor, and a very caustic and smoky taste. It is *soluble* in 60 to 90 parts of water, in all proportions of alcohol, ether, chloroform, benzole, petroleum naphtha, fixed and volatile oils, anhydrous glycerin, acetic acid, sulphuric acid (with combination and brown color), and in an equal part of bisulphide of carbon. It is soluble in aqueous alkalies—forming instable salts. It dissolves (and in commerce usually contains) about 3 per cent. of water, from which it is separated by mixture with a large quantity of benzole.

Creosote *resembles Phenic Acid*, in most of its physical properties, and in its reactions with nitric acid, ferric salts, bromine, gelatin, and albumen. It is *distinguished from Phenic acid* by not crystallizing when pure; by gelatinizing collodion; by not giving a blue color with ferric salts in a slightly alcoholic and sufficiently dilute solution of ferric chloride, as specified under Phenic acid, 35, *c* (Fluckiger's test); by not forming a clear mixture with a double volume of 18 to 20 per cent. ammonia, or with 5 volumes of ordinary (slightly aqueous) glycerin, or with a greater volume of bisulphide of carbon; and by more sparing solubility in water.

117. ANTHRACENE. $C_{14}H_{10}$. A colorless solid, crystallizing in the monoclinic system, often in four or six-sided tablets, having spec. grav. 1.147, melting at about 212° C., subliming slowly from the solid, and distilling rapidly at 300° C. When pure, the crystals show blue or violet fluorescence. It is tasteless and odorless, but its vapor at the distilling point is disagreeable and irritating.—It is insoluble in water, sparingly soluble in cold, moderately soluble in hot alcohol, soluble in ether, benzole, and oil of turpentine.—It is not affected by alkalies; is acted on by nitric acid, and dissolved with green color by sulphuric acid. With **picric acid**, in saturated alcoholic solution, it forms a salt crystallizing in red needles.

118. ALIZARIN. $C_{14}H_8O_4$. A yellow to red-yellow solid; by sublimation (at 215° C.) crystallizing anhydrous in red prisms, and from solutions crystallizing in golden scales of the hydrate.—Slightly *soluble* in water; soluble in alcohol and ether (with yellow color) and in concentrated sulphuric acid (with brown color); soluble in aqueous alkalies and alkaline carbonates (with purple color); these solutions being precipitated (orange) by acids, in good part even by carbonic acid gas. The ammoniacal solution, with salts of magnesium, iron, copper, and silver, forms purple and iridescent precipitates; the potassa solution is decolorized by lime-water, and the alcohol solution is decolorized by alumina with formation of a red precipitate.

119. BENZOLE. C_6H_5H with traces of its homologues. Coal-tar naphtha. Benzene.—A colorless limpid liquid, of about 0.85 spec. grav., crystallizing at 0° C., melting at 5.5° C., boiling at 80° or 81° C. (176° or 178° F.), and of a characteristic pleasant odor, reminding of rose and of chloroform. It burns with a bright, smoky flame. It is not perceptibly soluble in water (to which, however, it imparts odor), but is soluble in all proportions of alcohol, ether, chloroform, petroleum naphtha, etc. It dissolves sulphur, phosphorus, iodine, fixed and volatile oils, camphors; many resins (see 99, *f*); many alkaloids (not cinchonia) (133).

It is *distinguished* from Petroleum Naphtha by its generally greater solvent power (by dissolving hard pitch), and, more accurately, by its formation of *nitrobenzole* and products of the latter, as follows: Equal volumes of **nitric acid** of spec. grav. of 1.5 or of concentrated nitric acid containing nitrous acid, and of the liquid tested for benzole, are digested in a test-tube by immersion in hot water. The nitrobenzole rises in droplets, and is recognized by its odor of bitter almond oil and by its giving anilin with reducing agents, as stated at 120.

Or, for more delicate test—as in presence of Volatile Oils: A few drops of the liquid to be tested are mixed in a cooled tube with four times their volume of fuming nitric acid; the mixture is agitated and left a quarter of an hour; then mixed with ten times its bulk of water (which separates drops of nitrobenzole). Agitate with ether, which takes up the nitrobenzole;

decant the ether solution, filter, quickly distil the ether from the filtrate. To the residue add 1 or 2 c.c. of acetic acid and a particle of iron (filings), and distil over a very small flame. As soon as the liquid is nearly evaporated, add 2 or 3 c.c. of water and distil again. Mix the distillates (if acid, neutralize with slaked lime and filter), and test with chlorinated lime for anilin (violet color) (125, *a*).

119½. PETROLEUM NAPHTHA. Gasolene. "Benzene."—The rectified distillate of petroleum, having a boiling point of about 49° C. (120° F.)—specific gravity about 0.665. Consists chiefly of $\mathbf{C_5H_{11}H}$, with a little $\mathbf{C_6H_{13}H}$ and other homologues. *Characterized* by an agreeable odor and anæsthetic effect; by a wide range of solubilities; and by resisting the action of alkalies and most acids, while decomposed by heating with nitric acid.—*Distinguished from Benzole* by a lower specific gravity (even when both are of the same boiling point), and, more accurately, by not forming nitrobenzole (119).

120. NITROBENZOLE. $\mathbf{C_6H_5(NO_2)}$. "Essence of Mirbane." "Artificial oil of bitter almonds." Nitrobenzene.—A yellowish, oily liquid, of spec. grav. 1.21, crystallizing below 3° C., and boiling at 220° C. (428° F.) It has the odor of bitter almond oil, with equal persistence; a very sweet taste, and a highly poisonous effect taken by inhalation or through the mouth.—It is insoluble in water, freely *soluble* in alcohol, ether, chloroform, fixed and volatile oils.—It is *identified* by its odor—coinciding with its reaction for *anilin*. When a few drops are digested in a test-tube with zinc, acetic acid, and iron or magnesium wire, and the mixture extracted with ether, the residue of the latter gives reactions for anilin. See Benzole (119) and Anilin (121). Or a few drops are digested and shaken with zinc and dilute sulphuric acid, the mixture filtered through a wet filter, and the filtrate tested (with chlorate of potassium) for anilin. Both the above methods are applicable in presence of

Bitter Almond oil; also the following: Two or three cubic centimetres of the oil to be tested for nitrobenzole are agitated with about half its weight of fused potassa. If nitrobenzole is present, a reddish-yellow color appears, quickly turning to green, and if water is added there is separation of an upper layer of green, turning red the following day. Finely-divided zinc or iron, alone, digested at 100° C. for a day or two, reduces nitrobenzole to anilin.—Nitrobenzole is *distinguished* from bitter almond oil and other Volatile Oils by its specific gravity.

BASES: LIQUID AND SOLID.

121. ANILIN. $(C_6H_5)H_2N$. Monophenylamin.—Pure anilin is a colorless, limpid, oily liquid, of spec. grav. 1.028, vaporizing slightly at ordinary temperatures, boiling at 182° C. It is neutral to litmus, of bitter, burning taste, and vinous, aromatic odor. It is slightly *soluble* in water—the solution having a faint alkaline reaction; also it dissolves a little water. It is soluble in all proportions of alcohol, ether, chloroform, and most fixed and volatile oils, and in about equal volumes of bisulphide of carbon or benzole, but not perfectly in greater volumes of either. It is sparingly soluble in glycerin.

122. The Anilin Oil of Commerce contains more or less Toluidin, with traces of benzole, phenic acid, nitrobenzole, acetic acid, acetone, etc. "Kuphanilin" contains about 90 per cent. of phenylamin, and has a boiling point of 180° to 190° C. "Baranilin" is mostly toluidin, with a little cumidin and cymidin, boiling at 195° to 215° C.*

* (Mono)phenylamin, ... $(C_6H_5)H_2N$.
Toluidin, ... $(C_7H_7)H_2N$.
Xylilin, ... $(C_8H_9)H_2N$.
Cumidin, ... $(C_9H_{11})H_2N$.

123. *Phenylamin with acids forms salts,* crystallizable, soluble in water and in alcohol, many of them soluble in ether. The oxalate is sparingly soluble in cold absolute alcohol, insoluble in ether; the hydrochlorate is soluble in ether, not in cold chloroform. Anilin salts are readily decomposed by fixed alkali, when the anilin may be separated by ether. In the cold, anilin is displaced by ammonia; with heat, ammonia is displaced by anilin.

124. Toluidin has, with most solvents, nearly the same solubility as phenylamin. It forms few salts; the oxalate is sparingly soluble in water.

125. *Anilin is identified,* through formation of Anilin Red (rosanilin, fuchsin, or magenta), by chlorinated lime or chlorinated soda (*a*), by ferric chloride (*b*), by binoxide of manganese and sulphuric acid (*c*); and by its reaction with chlorate of potassium and hydrochloric or sulphuric acid (*d*), and with mercuric chloride (*e*). It is *distinguished from Alkaloids* (including Conia and Nicotia) by giving no precipitates with potassio mercuric iodide solution, or with iodine in iodide of potassium solution, or picric acid in presence of sulphuric. It *coincides with Alkaloids* in giving precipitates with phosphomolybdate (*f*), and with tannic acid (*g*).—It is characterized by a moderate *reducing power* (*h*). Anilin is examined as regards its *proportion of Toluidin,* as explained in *a*. It is separated from benzole, nitrobenzole, and other associated impurities by fractional distillation.

Anilin Red is a term for various salts and compounds of Rosanilin ($C_{20}H_{19}N_3 . H_2O$). This is a triatomic base which is colorless when pure—in the air becoming rose-red, or if formed in part from toluidin becoming brown, also dissolving freely in alcohol with a red color. It is nearly insoluble in water and insoluble in ether. It forms mono-acid salts having an intense crimson color (in solution), and tri-acid salts of yellowish-brown color.—As formed from commercial anilin, by oxidizing agents, rosanilin has a rich violet-purple color, changed to red by acids, and restored to violet-purple by alkalies. In proportion as formed from toluidin, the color becomes brown. Ether extracts the brown, leaving a blue.

a. A water solution of anilin or its salts, with a little solution of **chlorinated lime** or chlorinated soda gives a purple-red color, changing to brown-red by exposure to the air, or to rose-red by addition of acids. The color passes into a brown; if the mixture be shaken with ether, the latter rises to the surface as a brown layer, leaving a blue liquid below.

b. To a small portion (10 c.c.) of a very dilute solution of anilin, strongly acidulated with hydrochloric acid, add of a concentrated solution of **ferric chloride** two or three drops, or enough to give a yellowish tint, and heat gradually to boiling. The color becomes darker to opaque violet-brown. When a precipitate separates, filter and wash with water; then treat the precipitate with 60 per cent. alcohol, when the violet color is dissolved. The aqueous filtrate, shaken with chloroform, forms two light red layers.

c. A diluted solution of anilin, acidulated with sulphuric acid, on agitating with **binoxide of manganese,** quickly gives a blue to purple-red color, more intense after warming to 50° or 60° C.

d. **Chlorate of potassium** with hydrochloric or sulphuric acid, when strong, forms a red resinous substance; when dilute, a violet color.

e. **Mercuric chloride,** in the solid state, gently heated with anilin, converts it into a dark-purple mass which gives a red solution in alcohol.

f. **Phosphomolybdate of sodium** with solutions of anilin acidulated with sulphuric or oxalic acid, gives a blue precipitate becoming yellow (with Nicotia, the precipitate is yellowish at first). Addition of ammonia of 18 or 20 per cent. dissolves the anilin precipitate with deep blue color (the Conia precipitate is left blue but undissolved by ammonia).

g. **Tannic acid,** with solutions of anilin not very dilute and not containing free acid or free ammonia, a white precipitate of tannate of anilin.

h. Anilin reduces permanganate solution, but not potassio cupric sulphate.

126. ALKALOIDS.—Volatile and Non-volatile.—*The volatile alkaloids* are composed of **C, H,** and **N,** without **O**; and, in their consistence, vaporization, and other physical properties, resemble the volatile oils, but differ from them by approaching the character of ammonia. They are expelled from their salts by **fixed alkalies and heat.** The most important are the five following. (For Solubilities, see **133**; Separations, **134**; Comparative reactions, **131** and **135** to **143**.)

ANILIN (121).

127. CONIA.* $\mathbf{C_8H_{15}N}$. A colorless liquid, of spec. grav. 0.89, wasting slightly at ordinary temperatures, distilling almost wholly with steam at 100° C., boiling at 160° to 180° C. It has a mouse-like odor, sharp taste, and strong alkaline reaction. It resinifies, yellowish, in the air. Its administration causes enlargment of the pupil. It is a strong base; its salts being soluble in water and alcohol, not in ether. It coagulates albumen.

128. LOBELINA. An oily, volatile liquid, of alkaline reaction. Its administration dilates the pupils.

129. NICOTIA. $\mathbf{C_5H_7N}$. A transparent, oily liquid, of spec. grav. 1.048, distilling with steam at 100° C., or slowly alone at 146° C., boiling at 243° C. It has an ethereal, tobacco-like odor (when pure), and (in dilute solution!) an acrid taste. In reaction it is strongly alkaline. It resinifies in the air.

130. TRIMETHYLAMIA. $\mathbf{C_3H_9N}$. Propylamin. Secalin.—A colorless liquid below 5° C., its vaporizing point. (Soluble in water and alcohol.) It has an odor of herring and of ammonia, a sharp, bitter taste, and an alkaline reaction. Its salts are crystallizable, and soluble in water and (mostly) in alcohol. Its hydrochlorate is soluble in absolute alcohol (separation from Ammonia). Its water solution precipitates aluminum salts and then dissolves the precipitate (distinction from Ammonia). Its solution in equal weight of water is combustible.

* The termination *a* is given in this work to all the alkaloids, but the terminal *n* is used by many writers.

131. Comparative Reactions of the Volatile Bases. (F. F. Mayer.)

(W. *indicates water-solution ;* S., *salts or solution with acid.*)

	AMMONIA.	TRIMETHYLAMIA.	ANILIN.	NICOTIA.	CONIA.	LOBELINA.
Iodine in solution of KI.	W. Decolorized. S. No change.	W. & S. Orange-yellow precipitate.	Brown solution: precipit. forms slowly or not at all.	W. & S. Brown-red precipitate.	W. & S. Pale brown-red precipitate.	W. & S. Brown-red precipitate.
Tannic acid.	W. No precipitate. S. Precip. in conc. solution.	In neutral or alkaline solution, white, curdy precipitate.	No precipitate.	White precip., sol. in acids.	White precipitate, soluble in tannic and other acids.	White precipitate, sol. in ammonia and in tannic ac.
Mercuric chloride.	W. White precipitate.	W. White precipit.	No precipitate.	W. White precipit.	W. White precipit.	W. No precipitate.
Potassio Mercuric iodide.	S. No precipitate. W. Yel.-white precipitate, soluble in acids.	S. Pale yellow precipitate, cryst. and decomp. by much water. W. Precip., sol. in excess.	W. White precip., sol. in excess and in KI.	W. & S. Yellowish precip., not easily sol. in excess, sol. in potassa.	W. & S. The same as Nicotia.	W. & S. Pale yell'w precipit., slightly sol. in excess.
Subacetate of lead.	. . . . . .	. . . . . .	W. White precipitate.	. . . . . .	. . . . . .	. . . . . .
Nitrate of silver.	W. Brown precipit., soluble in ammonia.	W. Grayish precip., sol. in nitric acid.	No precipitate.	W. At first no precipit., then brownish-black precipit. (after warming).	W. White precipitate, turning brown, sol. in ammonia, at first sol. in nitric acid.	W. White precip., soluble in ammonia and in nitric acid.
Chloride of gold.	W. Reddish-yellow precipitate.	W. Gray-yellow precip., sol. in H Cl.	W. No precipitate.	W. Yellowish, curdy precip., insoluble in H Cl.	W. Whitish precip., insoluble in H Cl.	W. Pale yel. precipitate, insoluble in H Cl.
Platinic chloride.	. . . . . .	. . . . . .	. . . . .	Slight precip., soluble on warming.	S. Yellow; solub. in alcohol. (See 142.)	
Solubility in water, and spec. grav.	Very soluble.	Very soluble.	Slightly soluble; floats on water.	Sparingly soluble; sinks in water.	Sparingly sol.; floats on water.	Sparingly soluble; floats on water.

132. Non-volatile Alkaloids and accompanying Glucosides. (For Solubilities, 133; Separations, 134; Reactions common to alkaloids, 135, 142, 143.) (For Determinations of Quantity. 135*a*, 142, 143.)

ACONITA. $C_{30}H_{47}NO_7$.—Glacial mass or white powder. Crystallizes with difficulty.—136 (135, *e*, *f*).

ATROPIA. } $C_{17}H_{23}NO_3$.—Prisms; stellated tufts; white powder;
DATURIA. } fusible at 90° C.—136, 135.

BERBERINA. $C_{21}H_{19}NO_5(H_2O)_5$.—Light-yellow silky needles, or grouped prisms.—136, 138.

BRUCIA. $C_{23}H_{26}N_2O_4$.—Colorless; delicate needles; four-sided prisms.—136, 137, **138**, 140, 139.

CAFFEINA. $C_4H_5N_2O$.—White, silky needles; fusible at 178° C.; subliming at 185° C.—136, **140** (135, *a*, *e*, *g*).

CINCHONIA. $C_{20}H_{24}N_2O$.—Four-sided prisms or needles; fusible at 165° C.—136.

CINCHONIDIA. $C_{20}H_{24}N_2O$.—Hard rhombic prisms, with striated faces. Melts at 175° C.

CODEINA. $C_{18}H_{21}NO_3$.—Rectangular octahedrons; or (in presence of water) trimetric.—136, 138, 139.

COLCHICIA. $C_{17}H_{19}NO_5$.—Colorless prisms or needles; yellowish-white powder; glacial.—135, 136, 138, 140.

DAPHNIN. $C_{31}H_{38}O_{19}$.—Rectangular prisms. Odorous above 100° C.; above 200° C., Daphnetin.—138, 141.

DELPHINA.—Amorphous; powder white with yellow tint. Melts to resinous mass.—136 (135, *e*).

DIGITALIN. $C_{10}H_{18}O_4$.—Difficult to crystallize. A Glucoside.—136 (135*a*) (142).

EMETIA. $C_{30}H_{44}N_2O_4$.—Yellow-white powder. Melts at 50° C.—136, 138 (135, *e*).

ERGOTINA. $C_{50}H_{52}N_2O_3$.—Red-brown powder.—136.

HYDRASTIA.—Colorless, shining, four-sided prisms. Above 100° C., melts.—136, 137, 140.

HYOSCYAMIA. $C_{15}H_{23}NO_3$.—Stellate groups of silky needles; amorphous and pasty. Fusible.—136.

IGASURIA.—Colorless, lustrous prisms. Fusible.—136, 138, 140.

MORPHIA. $C_{17}H_{19}NO_3(H_2O)$. — Short, transparent, trimetric prisms. Anhydrous at 120°.—136, 138, **141**.

NARCEINA. $C_{23}H_{29}NO_9$.—Colorless. delicate needles. Fusible.—136, 137, 138, 139.

NARCOTINA. $C_{22}H_{23}NO_7$.—Colorless, rhombic prisms. Fusible.—136, 138, **139**.

OPIANIA. $C_{66}H_{72}N_4O_{21}$.—Right rhombic prisms.—138, **139**.

PAPAVERINA. $C_{20}H_{21}NO_4$.—Colorless, acicular crystals.—**136**, 138.

PAYTINA. $C_{21}H_{24}N_2O$.—Colorless crystals.

PHYSOSTIGMIA. $C_{15}H_{21}N_3O_2$. — Amorphous, brownish-yellow; solutions, red to blue.—136, **140**.

PICROTOXIN. $C_{12}H_{14}O_5$.—Needles; stellate; laminæ. Reduces cupric hydrate.—137.

PIPERIN. $C_{17}H_{19}NO_3$.—Colorless, monoclinic prisms. Melts at 100° C.—136, 138.

PSEUDOMORPHIA. $C_{17}H_{19}NO_4$.—Fine, lustrous crystals.—136, 138, 141.

QUINIA. $C_{20}H_{24}N_2O_2$.—Hydrate, in fine needles. Solutions, blue-fluorescent.—136, **140**.

QUINIDIA. $C_{20}H_{24}N_2O_2$.—Transparent, monoclinic prisms, efflorescent.—136, **140**.

RHŒADIA. $C_{21}H_{21}NO_6$.—Small, white prisms. Melts at 232° C.—Purple-red with acids.

SABADILLIA. $C_{20}H_{26}N_2O_5$.—Cubic crystals (Needles?). Reacts with sulph. acid like Veratria (136) 135, *e*.

SALICIN. $C_{13}H_{18}O_7$. — Tabular or scaly crystals. — Melts at 120° C. A Glucoside.—**136**.

SAPONIN. $C_{32}H_{54}O_{18}$.—Amorphous. Aromatic odor, sweet taste, burning after-taste. A Glucoside.

SOLANIA. $C_{43}H_{69}NO_{16}$.—Silky needles; right, four-sided prisms. A Glucoside.—136, 138.

STRYCHNIA. $C_{22}H_{24}N_2O_2$.—Four-sided prisms, trimetric, white. Fusible.—136, **137**.

THEBAINA. $C_{19}H_{21}NO_3$.—Thin, square tablets of silvery lustre. Fusible.—**136**, 138.

THEOBROMINA. $C_7H_8N_4O_2$.—Microscopic, trimetric crystals, in club-shaped groups.—136, **140**.

VERATRIA. $C_{32}H_{52}N_2O_8$.—White or greenish-white crystallized powder. Warmed with **HCl**, violet.—**136**.

133. Solubilities of the Alkaloids.—*In alcohol* they are generally freely soluble, the following being the only important exceptions and notices to be made:

Caffeina—in 30 parts strong alcohol.

Morphia—in 30 parts boiling or 50 parts cold absolute; in a somewhat smaller quantity of 90 p. c. alcohol.

Narceina—easily in hot, in 950 parts cold 85 p. c. alcohol.

Narcotina—in 25 parts boiling or 100 parts cold 85 p. c. alcohol.

Opiania—slightly in hot, scarcely at all in cold alcohol.

Pseudomorphia—nearly insoluble.

Solania—in 150 parts hot or 500 parts cold alcohol.

Strychnia—difficultly soluble in absolute, soluble in 115 parts of 95 p. c., 125 parts of 90 p. c., 130 parts cold or 15 parts boiling 75 p. c., 250 parts cold or 25 parts boiling 50 p. c. alcohol.

Theobromina—in 50 parts hot or 1500 cold alcohol.

The solubilities given for *ether* in the table refer to ether nearly or quite free from alcohol.

Benzole (of coal-tar), as used below, distils at 60° to 80° C. (140° to 176° F.), leaving *no* residue.

Amylic alcohol dissolves 0.1568 part of Codeina, 0.0026 part of Morphia, 0.0032 part of Narcotina, 0.0130 part of Papaverina, and 0.0167 part of Thebaina (KUBLY).

Ether dissolves from acid solutions—Colchicin, Digitalin, Picrotoxin—in general not the (other) alkaloids.

Petroleum Naphtha, as used below, distils at from 40° to 60° C. (104° to 140° F.), leaving no residue.

Amylic Alcohol should be strictly free from ethylic alcohol.

The acid used with chloroform, benzole, etc., is sulphuric acid, added just to an acid reaction, and forming sulphates of the alkaloids.

Glucosides have the termination N.

The * refers to explanation given below for the alkaloids, alphabetically.

	Water.	Fixed Alkali with water.	Ammonia with water.	Ether.	Chloroform.	Benzole.	Petroleum Naphtha.	Chloroform with acid.	Benzole with acid.	Petroleum Naph. with acid.	Amyl. Alc. with acid.
Aconitia.	Sol. 150 pts.	(As water)	Spar'g. sol.	Sol. 2 pts.	Sol. 2.5 pts.	Soluble.	Insoluble.	Insoluble.	Insoluble.	Insoluble.	Slight. sol.
Atropia.	In 60 pts. boil.	Soluble.	Soluble.	Sol. 30 pts.	Sol. 4 pts.	Sol. 50 pts.	Insoluble.	Insoluble.	Insoluble.	Insoluble.	Slight. sol.
Berberina.	Spar'g sol.	Soluble.	(As water)	Insoluble.	Slight. sol		Insoluble.		Slight. sol.	Insoluble.	Soluble.
Brucia.	In 500 pts. boil.	(As water)	Soluble.	Insoluble.	Sol. 4 pts.	Sol. 60 pts.	Sol. 120 pts.			Insoluble.	Slight. sol.
Caffeina.	In 90 pts. cold.	Soluble.	Soluble.	Sol. 500 pts.	Sol. 5 pts.	Soluble.	Insoluble.	Soluble.	Soluble.	Insoluble.	Soluble.
Cinchonia.	In 2500 pts. boil.	Insoluble.	Insoluble.	Sol. 400 pts.	Sol. 60 pts.	Soluble.	Near. insol			Insoluble.	
Cinchonidia.	In 2000 pts. cold.			Sol. 150 pts.						Insoluble.	
Codeina.	In 75 pts. cold.	(As water)	(As water)	Soluble.	Soluble.	Sol. 12 pts.	Slight. sol.	Insoluble.	Insoluble.	Insoluble.	Insoluble.
Colchicia.	Soluble.		Soluble.	Soluble.	Soluble.	Spar'g. sol.	Insoluble.	Soluble.	Soluble.	Insoluble.	Soluble.
Conia.	In 100 pts.*	Soluble.		Sol. 6 pts.	Soluble.	Soluble.	Soluble.			Insoluble.	
Daphnin.	Spar'g. sol.	Soluble.	Soluble.	Near. insol							
Delphina.	Insoluble.			Soluble.	Soluble.	Soluble.	Near. insol	Slight. sol.	Soluble.	Insoluble.	Soluble.
Digitalin.	Slight. sol.		Soluble.	Slight. sol.	Spar'g. sol.			Soluble.	Soluble.	Insoluble.	Soluble.
Emetia.	Spar'g. sol.	Soluble.	Spar'g. sol.	Near. ins'l.	Soluble.	Soluble.	Soluble.	Insoluble.	Insoluble.	Insoluble.	
Ergotina.	Soluble.			Insoluble.	Insoluble.						
Hydrastia.	Insoluble.			Spar'g. sol.	Soluble.						
Hyoscyamia.	Sol. hot.	(As water)	(As water)	Soluble.	Soluble.	Soluble.	Insoluble.	Insoluble.	Insoluble.	Insoluble.	Insoluble.
Igasuria.	Spar'g. sol.	Spar'g. sol.		Spar'g. sol.	Soluble.						
Lobelina.	Slight. sol.			Soluble.	Soluble.						
Morphia.	In 500, b'il.*	Soluble.	Slight. sol.	Insoluble*	Sol. 90 pts.	Insoluble.	Insoluble.	Insoluble.	Insoluble.	Insoluble.	Insoluble.
Narceina.	In 200, b'il.*	(As water)	Slight. sol.	Insoluble.	Spar'g. sol.	Slight. sol.	Insoluble.	Soluble.		Insoluble.	Soluble.
Narcotina.	In 7000, b'l.*	Insoluble.	Insoluble.	Sol. 120 pts*	Sol. 3 pts.	Sol. 25 pts.	Near. insol	Soluble.	Insoluble.	Insoluble.	Spar'g. sol.

	Water.	Fixed Alkali with water.	Ammonia with water.	Ether.	Chloroform.	Benzole.	Petroleum Naphtha.	Chloroform with acid.	Benzole with acid.	Petroleum Naph. with acid.	Amyl. Alc. with acid.
Nicotia.	Soluble.*	Soluble.		Soluble.	Spar'g. sol.	Soluble.	Soluble.	Insoluble.	Insoluble.	Insoluble.	Insoluble.
Opiana.	Slight. sol. hot.	(As water)	(As water)								
Papaverina.	Insoluble.	Insoluble.	Insoluble.	Slight. sol.	Soluble.	Sol. 40 pts.	Sol. warm.	Soluble.	Insoluble.	Insoluble.	Insoluble.
Paytina.	Slight. sol.			Soluble.	Soluble.		Soluble.				
Physostigmia	Slight. sol.	(As water)	Spar'g. sol.	Soluble.	Soluble.	Soluble.	Insoluble.	Insoluble.	Slight. sol.	Insoluble.	Spar'g. sol.
Picrotoxin.	Sol. 50 pts. hot.	Soluble.	Soluble.	Sol. 250 pts.	Soluble.		Soluble ?	Soluble.			Soluble.
Piperin.	Near. insol.	(Decomp.)		Sol. 90 pts.	Soluble.	Soluble.	Soluble.	Soluble.		Soluble.	Soluble.
Pseudomorphia.	Insoluble.	Soluble.	Insoluble.	Insoluble.	Insoluble.						
Quinia.	In 1800 pts.*	Insoluble.	Soluble.	Soluble.*	Sol. 50 pts.	Soluble.	Soluble.	Insoluble.	Insoluble.	Insoluble.	Insoluble.
Quinidia.	In 750 pts.			Sol. 30 pts.	Soluble.	Soluble.	Insoluble*	Insoluble.	Insoluble.	Insoluble.	Insoluble.
Rhœadia.	Insoluble.	Insoluble.	Insoluble.	Sol. 1300 pts	Spar'g. sol.	Spar'g. sol.					
Sabadillia.	Spar'g. sol.		Soluble.	Insoluble.							
Salicin.	Soluble.	Soluble.		Insoluble.		Near. insol			Near. insol		Soluble.
Saponin.	Soluble.	Soluble.	Soluble.	Insoluble.							
Solania.	In 8000, boil	Soluble.	Insoluble.	Sol. 4000 pts	Insoluble.	Slight. sol.	Insoluble.	Insoluble.	Insoluble.	Insoluble.	
Strychnia.	In 6500 pts.	Insoluble.	Spar'g. sol.	Insoluble*	Sol. 7 pts.	Sol. 160 pts.	Sol. 350 pts.	Insoluble.	Insoluble.	Insoluble.	Insoluble.
Thebaina.	Ins'l. (c'ld)	Insoluble.	Insoluble.	Soluble.	Spar'g. sol.	Sol. 18 pts.		Soluble.	Insoluble.	Insoluble.	Insoluble.
Theobromina	In 750 pts.	Soluble.	Soluble.	Near. insol	Spar'g. sol.	Slight. sol.	Insoluble.	Soluble.	Insoluble.	Insoluble.	Soluble.
Veratria.	In 1000 pts. hot.	(As water)	Spar'g. sol.	Sol. 12 pts.	Sol. 2 pts.	Soluble.	Slight. sol.	Slight. sol.	Slight. sol.	Insoluble.	Spar'g. sol.

* Conia—less soluble in hot *water* than cold (distinction from Nicotia).

* Morphia—nearly insoluble in *water* at 10° C., soluble in traces at 20° C. Sparingly soluble in *ether*, when both nas[illegible] nd amorphous.

* Narceina—soluble in 375 pts. *water* at 14° C.

* Narcotina—soluble in 25000 pts. cold *water*. Soluble in cold Acetic acid. Soluble in 50 pts. of boiling *ether*.

* Nicotia—soluble alike in cold or hot *water*.

* Quinia—*water* solubility in the table at 15° C.; at 20° C., 1650 parts; at 100° C., 1900 parts. It requires about 5[illegible] .te *ether* for solution. Of the alkalies, least soluble in soda solution; dissolving in 2200 parts lime-water.

* Strychnia—in absolute *ether*, insoluble; in ether of spec. grav. 0.725, soluble in 1800 parts; in ether of spec. [illegible] in 1200 parts. According to Wormley, soluble in 1400 parts of absolute ether.

134. Separation of Alkaloids from (solid) Albumenoid, Fatty, and Extractive Matters.—(1) The alkaloids are dissolved out, as salts (tartrates, sulphates, or acetates) by alcohol, heat; the filtered solution is evaporated to dryness, idue dissolved as before, etc. For removal from esidue is dissolved in slightly acidulated water; the (filtered) solution evaporated and the solution repeated, etc.—The residue, in which the alkaloid is a salt, is washed with ether, as long as the ether removes anything (OTTO'S modification, 1856). The washed residue is treated with alkali, in presence of **ether**, which dissolves the nascent alkaloid. The residue from the ether solution is, if necessary, purified by extraction with alcohol, or acidulated water, or each, as required (STAS' method, 1851). Also, this method is adapted for volatile as well as fixed alkaloids. [The extraction with ether may be followed by use of chloroform.]

(2) A somewhat more simple method upon the same principle, for non-volatile alkaloids only, with use of **chloroform** instead of ether, and with carbonization by **sulphuric acid** (RODGERS and GIRDWOOD, 1856). Designed, by its authors, for strychnia only; but applicable for all alkaloids soluble in chloroform and not decomposed by concentrated sulphuric acid at 100° C.

(3) The use of **amylic alcohol** (as in Otto's and Stas' method) to wash the *acid* solution of alkaloids clean of all matters soluble in amylic alcohol, and, after saturating with alkali, dissolving the alkaloid in the same solvent. Then, the amylic-alcohol-solution of alkaloids is washed with acidulated water, whereby the alkaloids are removed from the former solvent and taken up by the water as salts. This is repeated until purification is complete. The method is applicable only to those alkaloids not soluble in amylic alcohol from acid—see Table, 133 (USLAR and ERDMANN, 1861).

(4) The use of **animal charcoal** to withdraw an alkaloid (strychnia) from a solvent; after which the alkaloid is extracted

from the charcoal by a more effective solvent. (GRAHAM and HOFMANN, 1853.)

(5) **Dialysis** of the alkaloids, as salts, from colloid matters (GRAHAM, 1862).

(6) **Separation of Alkaloids from each other,** as well as from indeterminate matters, etc. The use of **petroleum naphtha, benzole, chloroform,** and **amylic alcohol,** each first in acid and then in alkaline solutions—extracting back to acidulated water (as in method of Uslar and Erdmann) when necessary to purify. Division of the alkaloids into about eight groups. (DRAGENDORFF, 1868.)

(7) Separation of alkaloids *from each other and from associated Glucosides,* by an extension of the method last named.

(8) Separation of alkaloids from each other by their solubility in **alkali.**

(9) Separation of pure alkaloids from each other by successive use of **ether, water,** and **chloroform.** (PRESCOTT.)

(1) *Otto's and Stas' Method.*—An aliquot part of the material is finely divided if solid, or concentrated if liquid, and subjected to digestion, at about 60° C. (140° F.), with a double weight of 90 per cent. **alcohol,** with addition of 0.5 to 2.0 grams **tartaric acid** (or oxalic acid). This extraction is completed by expression and digestion with another portion of alcohol, repeated two or three times.—The filtered liquid is now concentrated to a small bulk—by use of gentle heat, or in vacuum [or heat with partial vacuum from condensation by use of two connected flasks*], or by gentle heat in a stream of air with use of a tubulated retort. Fat is separated by filtration through a wet filter, and the filtrate evaporated nearly or quite to dryness, in vacuum or over sulphuric acid.—Macerate the residue in **absolute alcohol** (for 24 hours), and evaporate the filtrate at a heat not above 40° C. (104° F.)—Moisten the residue with water, and

* PRESCOTT : *Chem. News*, xx., p. 222 (1870, Jan)

add bicarbonate of sodium or potassium as long as there is effervescence.—Add three or four volumes of **ether** (free from oil of wine and not heavier than 0.725 s. g.) and agitate. Evaporate a portion of the clear ether upon a watch-glass; a residue in oily streaks, collecting into droplets, having a pungent odor and alkaline reaction, gives evidence of volatile alkaloids.

If *volatile alkaloids* are present, the material is farther treated for a short time with a little strong potassa solution, and then extracted in a flask or large test-tube with repeated portions of the ether. Acidulate the ether solution with dilute sulphuric acid, stopper tightly and agitate, and remove the ether layer. (Ammonia, anilin, nicotia, picolin, as sulphates, are not soluble in ether; conia sulphate is slightly soluble in ether.) Evaporate the ether, at ordinary temperature, and test for Conia.—To the acid watery residue add excess of potassa or soda, and extract with ether as before. Evaporate the ether at low temperature, and test the residue for volatile alkaloids (126).—For non-volatile alkaloids, unite this residue with any fixed residue left by that portion of ether taken after adding bicarbonate, and treat as follows:

For *non-volatile alkaloids*, evaporate the (first) ether-extract, dissolve in a very little very dilute sulphuric acid (leaving a decided acid reaction) and wash with *ether* (absolute or nearly so) as long as anything is washed away.—Now add fresh ether, then add excess of concentrated solution of carbonate of sodium or potassium and extract thoroughly with several portions of the ether.—The residue from the ether may be purified by extraction with *water acidified* by sulphuric acid; then the concentrated aqueous sulphate is treated with carbonate of potassium in excess and extracted with *absolute alcohol.*—At each evaporation the appearance of *crystals* is watched. Crystallization from ethereal solutions is greatly promoted by adding alcohol.

(2) *Rodgers and Girdwood's method.*—The material is digested with *dilute hydrochloric acid* at a moderate heat for about two hours; the filtered extract evaporated to dryness on the water-bath; the residue extracted with *water* and filtered;

and the filtrate supersaturated with *ammonia* and then extracted, in a flask, with **chloroform.**—The solid residue from the chloroform is moistened with **concentrated sulphuric acid** and left on the water-bath for half an hour or longer to carbonize foreign organic matters. When cool, it is then extracted with water.—The water solution is saturated with *ammonia* and again extracted with *chloroform.* If the residue of (a portion of) the chloroform blackens on warming with sulphuric acid, the (whole) residue is again treated with sulphuric acid on the water-bath, extracted with water, and the aqueous solution with ammonia and chloroform.

(3) *Method of Uslar and Erdmann.*—Digest the material, brought to the consistence of a thin paste and acidified with **hydrochloric acid,** for an hour or two, at 60° to 80° C. (140° to 176° F.), and strain and press through wet linen, washing the residue with water acidified with hydrochloric acid.—Evaporate the united solutions, with addition of clean **sand** and of **ammonia** in excess, and triturate to a powder. Boil the residue, repeatedly, with **amylic alcohol,** and filter the extracts hot through paper moistened with amylic alcohol. The filtrate is usually yellowish, and holds fatty and coloring matters, with the alkaloids, in solution.—Transfer the filtrate to a cylindrical vessel, add ten or twelve times its volume of water acidified with hydrochloric acid and nearly boiling, agitate vigorously and set aside. Remove the amylic-alcohol-layer with a pipette. (This should be nearly or quite free from all those alkaloids not soluble in amylic alcohol with acid—the only ones considered in this process. See Table, 133. This amylic alcohol, however, may well be washed with one portion of hot acidulated water.) Wash the water-acid liquid with several portions of amylic alcohol.—Concentrate the water-acid liquid, add ammonia in excess, and repeat the extraction with hot amylic alcohol.—If the liquid is colored, or if the residue of a few drops is blackened by a drop of sulphuric acid, again extract by much hot acidulated water, and then by amylic alcohol.

(4) *Method with Animal Charcoal.*—Shake two ounces animal charcoal in half a gallon of the aqueous, neutral or feebly acid, liquid; let the mixture stand 24 hours, with occasional shaking; filter; wash the charcoal once or twice with water; then boil half an hour with 8 ounces of alcohol of 80 to 90 per cent. (condensing and returning the evaporated alcohol.) Filter and evaporate the filtrate.—(Devised by its authors for separation of strychnia from beer.)

(5) *Dialysis.*—The aqueous liquid or suspended material is acidified with hydrochloric acid, and floated, in the dialyzer, over pure water. The dialyzed liquid usually contains foreign matter; still to be removed by some other process.

(6) *Use of Naphtha, Benzole, Chloroform, and Amylic Alcohol, each in Acid and in Alkaline Solutions. Dragendorff's Method.*—The finely-divided material is extracted several times with **water acidulated with sulphuric acid**, digesting several hours at a temperature of 40° to 50° C. (If it is desired to examine for the glucosides, colchicin, digitalin, solanin, the digestion should be made at ordinary temperatures. Piperin may be in part undissolved.) The filtrate is treated with sufficient calcined **magnesia** to leave only a slight acid reaction, and evaporated over a water-bath to the consistence of a syrup. This is placed in a flask, treated with three to four parts of 70 to 80 per cent. alcohol, acidulated with sulphuric acid, and digested with frequent agitation, for 24 hours, at about 30° C. When cold, the liquid is filtered, the residue being washed with alcohol. The filtrate is evaporated to remove all the alcohol, and diluted with water: solution **A.**

Solution **A** is digested and washed in a flask with **petroleum naphtha** (see 133), at about 35° C. Fats, colors, etc., and, if present, *Piperin* are dissolved: solution **B.**—The watery-acid residue from solution B is digested and washed with **benzole** at about 45° C. If a small portion of the decanted benzole gives a perceptible residue, the whole is nearly neutralized with **magnesia** or ammonia (leaving a distinctly acid reaction) and then thoroughly extracted with the benzole: solution **C.** This benzole solution is evaporated in glass dishes, for examination of the residue. It may contain *Caffeina, Colchicin, Cubebin, Delphina, Digitalin* (and traces of Berberin, Physostigmin, and Veratria). See also under (7), 134.—The watery-acid residue from solution C is now extracted with **amylic alcohol**: solution **D.** In this solution there may be *Berberin* (traces in C), *Narcotina* (perhaps only in part), *Physostigmia* (traces in C), *Theobromina, Veratria* (and traces of Aconitina and Atropina).—The watery-acid residue from solution D is now extracted with **chloroform**: forming solution **E.** This may contain *Papaverina, Narcotina* (if not wholly in D), *Thebaina*, and perhaps Veratria left from the benzole of C.

The watery-acid residue of E is now made slightly alkaline by **ammonia**, and extracted at about 35° C. with **petroleum naphtha.**—If a little portion of this solution gives a colored residue, the whole of it is thoroughly washed with much **water acidulated** with sulphuric acid, thus transferring the alkaloids to watery-acid solution, from which they are again extracted by making alkaline and washing with petroleum naphtha. This, solution **F,** in petroleum naphtha, may contain *Brucia, Conia, Emetia, Nicotia, Quinia, Strychnia,* and remaining traces of Veratria.—Two of these are volatile and liquid alkaloids, soluble from the residue in cold water. Evaporation of the naphtha leaves quinia and strychnia crystalline; brucia, emetia, and veratria, amorphous. Quinia, emetia, and veratria are soluble in absolute ether; brucia and strychnia insoluble.

The watery-alkaline residue of F is now extracted several times with **benzole,** at about 40° C. If a portion of the benzole leaves a colored residue, the whole is extracted with acid-water and again taken up with benzole for purification, as directed above for naphtha. Solution **G** (in benzole) contains *Aconitia, Atropia, Cinchonia, Codeina, Hyoscyamia, Physostigmia, Quinidia.*—These alkaloids are all soluble in ether, except cinchonia. If the residue of the others is dissolved in water acidulated with sulphuric acid and then supersaturated with ammonia, aconitia and quinidia are precipitated; while atropia, codeina, hyoscyamia, and physostigmia are (for a brief time) dissolved. The aconitia and quinidia precipitate being dissolved in hydrochloric acid, platinic chloride precipitates only the quinidia.

The watery alkaline residue of G is now acidulated with sulphuric acid and washed at about 55° C. with amylic alcohol; then made alkaline with ammonia and extracted with **amylic alcohol** at the same temperature. If the residue of a portion of the solvent is colored, the alkaloids are extracted from the whole by acidulated water; and again extracted by amylic alcohol in presence of alkali (as directed for F and G). Solution **H** (in amylic alcohol) contains *Morphia, Narceina, Solania.*—Narceina is dissolved from the residue by warm water.

The watery alkaline residue of H, which may be termed solution **I**, may contain *Curarin,* and traces of Berberina (Digitalin) and Narceina. The solution (I) evaporated to dryness, with pulverized glass, yields its alkaloids to **alcohol.**

(7) *Separation of Alkaloids and Glucosides from each other. Dragendorff's scheme.* Use of the solvents and operations employed in (6).

A. **Benzole** *dissolves,* from *acid* (sulphuric) aqueous solutions —Caffeina, Colchicin (incompletely), Colocynthin, Cubebin, Delphina (incompletely), Digitalin, Elaterin, Narceina, Piperin, Syringin—and traces of physostigmia and veratria.

B. Benzole *dissolves,* from *alkaline* (ammoniacal), aqueous solutions—Aconitia, Atropia, Brucia, Cinchonia, Codeina, Conia, Delphina, Emetia, Hyoscyamia, Narceina (imperfectly), Narcotina, Nicotia, Papaverina, Physostigmia, Quinia, Quinidia, Strychnia, Thebaina, Veratria.

C. Benzole *fails to dissolve,* more than traces, from alkaline solutions—Morphia, Salicin, Solania, Syringin, Theobromina.

D. Benzole does *not dissolve,* either from acid or alkaline water solutions—Curarin, Picrotoxin, Salicin, Theobromina.

E. Benzole, Petroleum Naphtha, Amylic Alcohol, and Chloroform, all *fail to dissolve,* from acid or alkaline solutions, Curarin.

F. **Amylic alcohol** *dissolves,* from acid (sulphuric) water solutions, more readily when warm—Aconitia (very sparingly), Berberina (in greater part), Brucia (in traces), Caffeina, Cantharidin, Colchicin, Cubebin, Delphina, Digitalin, Narceina (sparingly), Narcotina, Picrotoxin, Piperin, Salicin, Santonin, Theobromina, Veratria.

G. **Petroleum Naphtha** *leaves undissolved,* from acid or alkaline solutions—Aconitia, Berberina, Caffeina, Curarin, Narceina, Salicin, Syringin, Physostigmia, Theobromin.

H. Petroleum Naphtha *dissolves* from *acid* (sulphuric) watery solutions—Piperin, Populin.

I. Petroleum Naphtha *dissolves* from *alkaline* (ammoniacal) solutions—Brucia, Conia, Emetia, Nicotia, Papaverina, Quinia, Strychnia, Veratria, and *traces* of aconita, berberina, cinchonia, delphina, narcotina.

J. Petroleum Naphtha does *not dissolve,* from alkaline solution—Caffeina, Colchicin, Delphina.

K. **Chloroform** *dissolves* from the *acid* (sulphuric) water solution—Caffeina, Colchicia, Colocynthin, Cubebin, Delphina (sparingly), Digitalin, Narcotina, Papaverina, Piperin, Picrotoxin, Santonin, Thebaina, Theobromina.

L. Chloroform *dissolves* from the *alkaline* (ammoniacal) water solution—Aconitia, Atropia, Berberina (sparingly), Brucia, Caffeina, Cinchonia, Codeina, Colchicin, Conia, Cubebin, Delphina, Digitalin, Emetia, Hyoscyamia, Morphia (sparingly), Narcotina, Narceina (sparingly), Nicotia, Papaverina, Piperin, Quinia, Strychnia, Thebaina, Theobromina, Veratria.

(8) *Separation of certain alkaloids from each other by solubility in alkali.*

A. Solutions of **Fixed Alkalies** precipitate, and by excess redissolve, in dilute solution—Atropia, Berberina, Codeina,

Conia, Hyoscyamia (partly), Morphia, Nicotia, Solania—Colchicin being decomposed.

Most of the other well-known alkaloids are left in precipitate by excess of fixed alkali.

B. Of those not redissolved by fixed alkalies, **Ammonia** in strong excess dissolves from precipitate—Aconitia, Colchicin, Hyoscyamia, Physostigmia, Strychnia.

(9) *Separation by successive use of Ether, Water (and Chloroform).*—The alkaloids are previously obtained pure, as bases, and in the solid state finely divided. The *ether* used is absolute, applied in proportion of 40 to 60 parts to one of the solid, with agitation and digestion in a stoppered flask. The *water* is applied hot, and in proportion of fully 100 parts to one of solid. The *chloroform* should be nearly or quite free from alcohol, and 20 to 40 parts used. Alkaloids which are appreciably divided by the solvents have their names placed in parentheses.

ALKALOIDS.

Treat with **Ether** *and filter.*

Evaporate Filtrate (a) to residue (a). *Treat* residue (a) *with* **water.**		Residue (b). *Treat* Residue (b) *with* **water.**			
Filt. (A) *Evap.* to res. (A).	Res. (B).	Filtrate (c). *Evap.* to residue (c). *Treat* (c) *with* **chloroform.**		Residue (D) *Treat* (D) *with* **chloroform.**	
		Filt. (C). *Evap.* to resid. (C)	Res. (D).	Filt. (E) *Evap.* to res. (E).	Res. (F).
(A)	(B)	(C)	(D)	(E)	(F)
Sol. in Ether.	*Sol. in Ether.*	*Ins. in Ether.*	*Ins. in Ether.*	*Ins. in Ether.*	*Ins. in Ether.*
" *Water.*	*Ins. in Water.*	*Sol. in Water.*	*Sol. in Water.*	" *Water.*	" *Water.*
		" *Chl'm.*	*Ins. in Chl'm.*	*Sol. in Chl'm.*	" *Chl'm.*
Aconitia.	Delphina.	(Berberina).	(Berberina).	Cinchonia.	(Digitalin.)
Atropia.	(Hydrastia).	Brucia.	Ergotina.	(Digitalin).	Pseudomorphia.
Codeina.	(Lobelina).	Caffeina.	(Narceina).	(Hydrastia).	Solania.
Colchicin.	(Narcotina).	Emetia.	(Salicin).	Morphia.	(Theobromina).
Conia.	Paytina.	(Igasuria).	(Saponin).	(Narceina).	
Hyoscyamia.	Physostigmia.	(Narceina).		Papaverina.	
(Igasuria).	Piperin.	(Picrotoxin).		Rhœadia.	
Nicotia.	Quinia.	(Salicin).		Strychnia.	
(Picrotoxin).	Quinidia.	(Saponin).		(Theobromina)	
	Thebaina.				
	Veratria.				

135. Detection and separation of alkaloids as a class. The material is obtained in solution, and free from albumenoid, gelatinoid, gummy, coloring, and "extractive" substances. Also, the presence of inorganic acids, bases, or salts which react with the several reagents, must be avoided. Then, *the alkaloids are precipitated*—by potassio mercuric iodide (also a means of volumetric determination) (*a*); by phosphomolybdic acid (permitting a division of alkaloids by subsequent use of ammonia) (*b*); by metatungstic acid (*c*); by potassio cadmic iodide (*d*); by picric acid (with distinguishing exceptions) (*e*); by tannic acid (with exceptions and peculiarities) (*f*); by solution of iodine with iodide (*g*).

a. The **potassio mercuric iodide** reagent is prepared by adding to solution of mercuric chloride enough potassic iodide to dissolve the precipitate first formed. It gives precipitates in even dilute solutions of nearly *all alkaloids except Caffeina, Colchicin, Digitalin, Theobromina;* the precipitates being mostly yellowish-white. For the reactions with the Volatile Alkaloids and Ammonia, see **131**. The precipitates are insoluble in acids (distinction from ammonia), or in dilute alkalies, but soluble in alcohol and (in many cases) in ether—also, in many cases, soluble in excess of the precipitant.—For the *extraction of the alkaloid* from the precipitate, triturate the latter with stannous chloride and enough potassa solution to give a strong alkaline reaction, then exhaust with ether or chloroform, or, if the alkaloid is not soluble in these, add potassic carbonate instead of potassa and extract with strong alcohol.

For the *volumetric determination* by potassio mercuric iodide (MAYER), the reagent is prepared with 13.55 grams mercuric chloride, 5 grams potassic iodide, and water to one litre. Of this standard solution, 1 c.c. precipitates, of each alkaloid, the quantities stated below:

Aconitia,	0.0268 gram.	Cinchonia,	0.0102 gram.
Atropia,	0.0145 "	Conia,	0.0042 "
Brucia,	0.0233 "	Morphia,	0.0200 "

Narcotina,	0.0213 gram.	Quinidia,	0.0120 gram.
Nicotia,	0.0040 "	Strychnia,	0.0167 "
Quinia,	0.0108 "	Veratria,	0.0269 "

The volumetric determination is somewhat unsatisfactory, by reason of the slowness with which the precipitate subsides. The alkaloid solution is slightly acidulated with sulphuric or hydrochloric acid ; after each addition of the reagent the mixture is strongly shaken and left to subside ; then a drop of the clear liquid is placed on a blue or black glass plate, and treated with a drop of the reagent—to learn whether further addition is necessary.

b. **Phosphomolybdic acid** solution *—SONNENSCHEIN'S Reagent—gives amorphous and mostly yellow precipitates with the alkaloids, as below. The alkaloid solution should be neutral or slightly acid, as alkalies dissolve the precipitate in most cases. The reaction with ammonia should be noted ten minutes after its addition.

	PRECIPITATE.	WITH AMMONIA.	ON BOILING.
Aconitia.	Yellow.	Blue solution.	Colorless.
Anilin.	Blue, then yellow.	" "	
Atropia.	Yellow.	Blue to colorless sol.	Colorless.
Berberina.	"	Blue solution.	"
Brucia.	Orange.	Yellow-green solution.	Brown.
Caffeina.	Yellow.	Colorless solution.	
Cinchonia.	Whitish-yellow.	" "	
Codeina.	Brownish-yellow.	Green solution.	Orange-red.
Colchicin.	Yellow.	Bluish solution, in ½ hr. greenish.	
Conia.	Yellow-white.	Bluish or greenish pre.	Colorless.

* The yellow precipitate formed on mixing acid solutions of molybdate of ammonium and phosphate of sodium—the phosphomolybdate of ammonium—is well washed, suspended in water, and heated with carbonate of sodium until completely dissolved. The solution is evaporated to dryness, and the residue gently ignited till all ammonia is expelled (sodium being substituted for ammonium). If blackening occurs, from reduction of molybdenum, the residue is moistened with nitric acid and heated again. It is then dissolved with water and nitric acid to strong acidulation ; the solution being made ten parts to one of residue. It must be preserved from contact with vapor of ammonia.

	PRECIPITATE.	WITH AMMONIA.	ON BOILING.
Delphina.	Gray-yellow.		
Digitalin.	Yellow, on warming dissolv's gr'n.	Blue solution.	Green, then colorless.
Emetia.	Yellowish.		
Ergotina.	(A precipitate.)		
Morphia.	Yellowish.	Dark blue sol., in ½ hr. a blue residue falls.	
Narceina.	Brown-yellow, becoming resinous.		
Narcotina.	Brown-yellow.		
Nicotia.	Yellow.	Blue solution.	
Papaverina.	In dilute sol., no precipitate.		
Physostigmia.	Yellow.	Blue precipitate.	
Piperin.	Brown-yellow.	Colorless solution.	
(Piperidin.	Clear yellow.	Blue solution.)	
Quinia.	Yellow-white.	Whitish precipitate.	
Quinidia.	"	" "	
Solania.	Yellow.	Colorless solution.	
Strychnia.	Yellow-white.	" "	
Theobromina.	"		
Veratria.	Yellow.	Colorless precipitate.	

c. **Metatungstic acid** precipitates alkaloids from very dilute solutions (SCHEIBLER). The reagent may be prepared by adding phosphoric acid to a solution of ordinary tungstate of sodium as long as a precipitate is formed and redissolved. The precipitates are white and flocculent. This test is more delicate than that with phosphomolybdic acid. Scheibler states that a distinct turbidity is produced in a solution of one part of quinia or strychnia in 200,000 of water.

d. **Potassio cadmic iodide** solution (prepared like potassio mercuric iodide*) (MARME's test) gives gray-yellow to yellow precipitates with the alkaloids. The solution of alkaloid should be feebly acidulated with sulphuric acid. The precipitates are easily soluble in alcohol, insoluble in ether, soluble in excess of the reagent, and decompose on long standing. Precipitates are obtained with

* Dissolve 20 parts iodide of cadmium and 40 parts iodide of potassium in 120 parts of water.

Aconitia,	Delphina,	Piperin,
Atropia,	Emetia,	Piperidin,
Berberina,	Hyoscyamia,	Quinia,
Brucia,	Morphia,	Quinidia,
Cinchonia,	Narceina,	Sanguinarin (red),
Codeina,	Narcotina,	Strychnia,
Conia,	Nicotia,	Thebaina,
Curarin,	Papaverina,	Veratria.
Cytisin,	Physostigmia,	

No precipitates are obtained (in dilute solutions) from Colchicin, Solania, Theobromina, or from other known glucosides and neutral substances.—The alkaloids are obtained from their precipitates by adding an excess of carbonate of sodium, drying, and extracting with ether, chloroform, or benzole, according to the solubility of the alkaloids sought.

e. **Picric or Trinitrophenic acid** precipitates from water solutions the larger number of the alkaloids, especially as sulphates. Presence of free sulphuric acid generally promotes these precipitations and enables them to be formed in more dilute solutions. On the contrary, they are dissolved by hydrochloric acid.

No precipitates are formed by picric acid, in acid sulphate solutions of Anilin, Caffeina, Morphia, Pseudomorphia, Solania (unless by long standing), Theobromina, and the Glucosides.—Aconitia and Atropia are not precipitated except in concentrated solutions.—Atropia and Morphia are, however, precipitated in neutral solutions.—*Sabadillia* in 150 parts of water is not precipitated.

Full precipitates are obtained from the strongly acid sulphates of Berberina, Colchicin, Delphina, Emetia, the Cinchona alkaloids, the Opium alkaloids with the exceptions above given, the Strychnos alkaloids, Veratria, etc.

The following results are obtained by treating about a grain of a water solution of (neutral) salt of the alkaloids with an *alcoholic solution* of picric acid (Wormley):

	Precipitate.	Least quantity of alkaloid showing precipitate.
Aconitia.	Yellow, amorphous.	$\frac{1}{20000}$ grain.
Atropia.	Yellow, crystalline.	$\frac{1}{5000}$ "
Brucia.	Yellow.	$\frac{1}{40000}$ "
Codeina.	Yellow, amorphous.	$\frac{1}{2500}$ "
Conia.	Yellow, crystalline.	$\frac{1}{1000}$ "
Morphia.	Yellow, amorphous.	$\frac{1}{500}$ "
Narceina.	" "	$\frac{1}{500}$ "
Narcotina.	" "	$\frac{1}{20000}$ "
Nicotia.	" "	$\frac{1}{50000}$ "
Solania.	" "	$\frac{1}{1000}$ "
Strychnia.	Yellow, crystalline.	$\frac{1}{25000}$ "
Veratria.	Yellow, amorphous.	$\frac{1}{5000}$ "

The alkaloids may be *extracted from their picrates* by addition of an alkali and chloroform, benzole, or other suitable solvent. (Alcohol does not dissolve potassic picrate; but it takes up the excess of potassa.)

f. **Tannic acid**—in solution with 8 parts of water and 1 part of alcohol—gives whitish, grayish-white, or yellowish precipitates with nearly all the alkaloids. In the larger number of instances these precipitates are easily soluble in acids, frequently dissolving in excess of the tannic acid; on the contrary, some of the alkaloids are precipitated by tannic acid only in strong acid solutions. Ammonia dissolves the tannates of the alkaloids.

No precipitates are obtained with Piperin, Salicin, or Saponin.

Dilute *acetic acid dissolves* the precipitates of tannates of Aconitia, Brucia, Caffeina, Colchicin, Morphia, Physostigmia, Quinia (if the acid is not very dilute), Solania, Veratria.

Cold dilute *hydrochloric acid does not dissolve* the precipitates of tannates of Aconitia, Berberina, Brucia (slightly dissolves), Caffeina, Cinchonia, Colchicin (dissolves slightly), Delphina, Digitalin, Narcotina, Papaverina, Thebaina, Solania, Strychnia (dissolves sparingly), Veratria.

Cold dilute *sulphuric acid does not dissolve* the precipitates of tannates of Aconitia, Physostigmia, Quinia, Solania, Veratria.

Precipitates are formed in neutral solutions (not very dilute), but not in slightly acid solutions, yet completely formed *in solutions strongly acidulated with sulphuric acid,* by Aconitia, Physostigmia, Solania, Veratria.

Concerning the reactions of the Volatile Alkaloids with tannic acid, see 131.

Alkaloids are *separated from their tannates* by mixing the moist precipitate with oxide or carbonate of lead, drying the mixture, and extracting with alcohol, ether, or chloroform.

g. Water solution of **iodine in iodide of potassium** precipitates the alkaloids in general. The solution is made of 3 parts of iodine, 5 of iodide, and 50 of water. (WORMLEY: 1 of iodine, 3 of iodide, and 60 of water.)—The precipitates are yellow, orange-yellow, reddish-brown, and brown.—*No precipitates* are obtained with (Ammonia), Caffeina (in neutral solution), Digitalin (or but slight turbidness), Solania, Theobromina.—*Yellow* precipitates are given by Atropia (sparingly saturated), Hyoscyamia, Physostigmia, and Trimethylamia (orange-yellow). *Red-brown* precipitates are obtained with Aconitia, Codeina, Conia, Lobelina, Morphia, Narceina, Narcotina, Nicotia, Quinia, Strychnia, and Veratria.

136. Concentrated sulphuric acid gives characteristic reactions with some of the alkaloids; and a greater number of good indications are given by FRŒHDE's reagent, which consists of 0.01 gram **molybdate of sodium** dissolved in 10 c.c. of concentrated sulphuric acid (and so prepared freshly each time it is required). For these tests the alkaloids must be almost absolutely free from impurities not alkaloids. One or two miligrams of the alkaloid are dropped upon 15 drops of the acid.

	CONC. SULPHURIC ACID.	FRŒHDE'S REAGENT.
Aconitia.	Slight yellow to yel.-br'n.	Yellow-brown ; colorless.
Amygdalin.	Light violet-red.	
Atropia.	Colorless solution.	Colorless.

	CONC. SULPHURIC ACID.	FRŒHDE'S REAGENT.
Berberina.	Dark olive-green.	Greenish-brown to brown.
Brucia.	Pale rose.	Red ; yellow.
Caffeina.	Colorless.	Colorless.
Cinchonia.	Colorless.	Colorless.
Codeina.	Colorless.	Green ; blue ; yellowish.
Colchicin.	Yellow.	Yellow.
Colocynthin.		Cherry-red (slowly).
Colombin.	Orange, turning red.	
Conia.	Colorless (*pale* reddish ?).	Pale yellow.
Cubebin.	Bright red, then crimson.	
Curarin.	Lasting blue.	
Delphina.	Brownish.	Red-brown.
Digitalin.	Brown to red-brown.	Orange ; cherry-red ; br'wn.
Elaterin.	Red.	Yellow.
Emotia.	Brownish.	
Ergotina.	Red-brown.	
Hesperidin.	Yellow-red.	
Hydrastia.	Colorless ; after heating, purple.	
Hyoscyamia.	Brownish.	
Igasuria.	Rose-color : yellowish ; greenish.	
Limonin.	Yellow-red.	
Meconin (Opianyl).	(With heat, blue to purple).	
Morphia.	Colorless.	Violet ; green-yellow ; violet.
Narceina.	Brown to yellow.	Yellow-brown ; yellowish ; colorless.
Narcotina.	Yel. ; purple after warm'g.	Green ; yellow ; reddish.
Nicotia.	Colorless.	Yellowish ; reddish.
Ononin.		Red.
Papaverina.	Violet ; blue.	Violet ; blue ; yel'w ; colorless.
Phloridzin.		(Slowly) blue.
Physostigmia.	Yellow ; olive-green.	
Piperin.	Pale yellow ; brown.	Yellow ; brown.
Populin.	Red.	Violet.
Pseudomorphia.	Olive-green.	
Quinia.	Colorless.	Colorless ; greenish.
Quinidia.	Nearly colorless.	Colorless ; greenish.
Salicin.	Bright red.	Violet ; cherry-red.
Sarsaparillin.	Deep red, then violet, then yellow.	
Senagin.	Yellow-red.	

	CONC. SULPHURIC ACID.	FRŒHDE'S REAGENT.
Smilacin.	Yellow-red.	
Solania.	Reddish-yellow.	Cherry-red ; red-brown ; yellow.
Strychnia.	Colorless.	Colorless.
Syringin.	Blood-red, then violet-red.	
Tannic acid.	Purple-red.	
Thebaina.	Blood-red ; yellow-brown.	Orange.
Theobromina.	Colorless.	Colorless.
Veratria.	Slowly to crimson red.	Yellow ; cherry-red.

137. Sulphuric acid and bichromate of potassium: the solid alkaloid being dissolved in the acid and a very minute fragment of the bichromate being brought into contact with the liquid.

With *Strychnia*, a brilliant play of changing colors, blue turning soon to violet and then red-violet, then slowly fading —(delicate and distinctive). With Brucia, an orange or brownish-orange color. With Narceina, a dirty-red mixture. With Hydrastia, a brick-red to carmine-red color; with Picrotoxin, red-brown. With anilin, a yellowish to greenish tint first appears, slowly passing into blue, which after half an hour or longer becomes nearly or quite black. With Curarin, a play of colors similar to strychnia (compare 136). With aconitia, atropia, codeina, conia, morphia, narcotina, nicotia, solania, veratria, and many other alkaloids,—there is only the slowly formed greenish color of chromic oxide.

This, *the strychnia test,* may be made with substitution of other oxidizing agents for the bichromate, the crystallized **permanganate** of potassium perhaps giving the best results. SONNENSCHEIN advocates the use of ceroso-ceric oxide.

138. Concentrated Nitric acid, of spec. grav. 1.42, gives a red or reddish-yellow color with the greater number of the alkaloids.

Brucia, in the solid state, is dissolved by nitric acid with intense blood-red color—solutions of the alkaloid giving the same with less intensity or a reddish-yellow color. On warming, or standing, the color changes to yellow: if now a drop of solution of **stannous chloride** is added, a purple color appears. The purple is discharged by either nitric acid or excess of stannous salt. *Igasuria* gives nearly the same reaction, both with nitric acid and stannous chloride; the violet to purple color with the last-named reagent being characteristic of brucia and igasuria.

Morphia in somewhat concentrated solutions is colored yellow to orange-red—the color is either not changed or is altered toward the yellow by stannous chloride (distinction from brucia).—*Codeina, Narceina,* and *Papaverina* are colored red to orange-yellow by nitric acid; and *Narcotina, Pseudomorphia, Opiania, Thebaina,* and *Rhœadia,* yellow. *Emetia* is changed to a yellow, resinous mass, with partial decomposition.

Colchicin is colored violet by nitric acid: the most concentrated nitric acid, containing nitrous acid, forming an intense blue-violet color. The color changes to brown, and finally to yellow—these tints being more distinct in proportion as the violet is deeper. If the chloroform solution of colchicin is treated with concentrated nitric acid, a violet-red color is formed and taken up by the chloroform layer.—*Curarin* is colored purple by nitric acid.

Nitric acid *produces no color* with Atropia (brown tint, fading), Caffeina, Cinchonia, Conia (sometimes yellowish), Quinia, Quinidia, Solania (becoming faint rose-red with bluish rim), Theobromina.

Berberina is colored brown by nitric acid.

Daphnin is colored red.

Piperin becomes greenish-yellow, orange, then red, and resinous.

139. Concentrated **sulphuric acid followed by nitrate of potassium** (solid), with *Narcotina* gives a deep blood-red color (delicate and distinguishing). The color is discharged by much excess of nitric acid.—In the same test, *Brucia* gives an orange-red, and *Opiania* a scarlet-orange color. *Codeina* becomes first greenish, then reddish. *Narceina* turns reddish-brown.

140. Chlorine water followed by ammonia.—*Quinia* (or

Quinidia) treated first with fresh chlorine water and then with ammonia, gives a green flocculent precipitate which by excess of ammonia dissolves to an emerald-green solution (characteristic). On neutralization with an acid, the color changes to light blue, which becomes violet or red on supersaturation with acid, returning to green with addition of excess of ammonia. Addition of solution of red ferricyanide of potassium to the ammoniacal green solution produces a red color (with *Quinidia* a bulky precipitate). A better result is obtained by adding the ferricyanide after the chlorine and before the ammonia. The impure chlorine obtained by addition of hydrochloric acid to chlorate of potassium serves the purpose of this test.

Colchicin, when treated with chlorine and ammonia, gives an orange solution.—*Caffeina* and *Theobromina*, treated with chlorine water (or nitric acid), then evaporated to dryness, on addition of ammonia give a purple-red color. *Chlorine*, alone, with *Brucia* and with *Igasuria* gives a light red color; with *Hydrastia*, blue fluorescence. *Physostigmia*, with solution of chlorinated lime, gives an intense red color, turning nearly black by farther addition.

141. Solution of **Ferric chloride** (dilute) colors solid *Morphia*, and *Pseudomorphia* blue. Also *Daphnin* blue in the cold, turning yellow when warmed.

Morphia separates iodine from **iodic acid.**

142. Platinic chloride solution precipitates the greater number of the alkaloids, even dilute solutions (those in 2,000 or 3,000 parts of water)—the precipitates being yellow, whitish-yellow or grayish-yellow, and some of them being soluble in cold hydrochloric acid.—Anilin, Digitalin, Physostigmia, and Solania, are *not precipitated;* and Aconitia, Atropia, Codeina, Hyoscyamia, Narcotia, Nicotia, Sabadillia, and Veratria only from concentrated solutions.—The alkaloids next named give precipitates; each precipitate, after *ignition*, leaving a weight of pure platinum bearing a fixed ratio to the weight of the alkaloid—in accordance with the formula given.

	P. C. Pt in precip	Color, etc.	Solubility in cold HCl.
Berberina, $(C_{20}H_{17}NO_4.HCl)_2PtCl_4$	18.1	Yellow, needles.	Soluble.
Brucia, $(C_{23}H_{26}N_2O_4.HCl)_2PtCl_4$	16.5	(Like Strychnia.)	
Caffeina, $(C_8H_{10}N_4O_2.HCl)_2PtCl_4$	24.5	Orange-yellow, granular.	Insoluble.
Cinchonia, $C_{20}H_{24}N_2O(HCl)_2PtCl_4$	27.4	Light yellow.	Insoluble.
Cinchonidia, $C_{20}H_{24}N_2O(HCl)_2PtCl_4$	27.4	Pale orange.	
Codeina, $(C_{18}H_{21}NO_3.HCl)_2PtCl_4$	19.2	Yellow.	
Colchicin,	.	(Like Morphia.)	
Conia, $(C_8H_{15}N.HCl)_2PtCl_4$	29.4	(Dissolves in alcohol, yellow.)	
Delphina,	17.4	Gray yellow, flocculent.	Soluble.
Emetia,	.	Yellow-white.	
Hyoscyamia,	.	Brownish, flocculent.	
Morphia, $(C_{17}H_{19}NO_3.HCl)_2PtCl_4$	19.5	Yellow, curdy; after 24 hours, crystalline.	Insoluble.
Narceina, $(C_{23}H_{29}NO_9.HCl)_2PtCl_4$	14.6	Yellow, crystallizable.	
Narcotina, $(C_{22}H_{23}NO_7.HCl)_2PtCl_4$	15.9	Yellow.	
Nicotia, $C_{10}H_{14}N(HCl)_2PtCl_4$	34.2	Orange-yellow (see 131).	Soluble.
Papaverina, $(C_{20}H_{21}NO_4.HCl)_2PtCl_4$	17.8	Yellow-white.	Soluble.
Quinia, $C_{20}H_{24}N_2O_2(HCl)_2PtCl_4$, dried at 100° C.	26.8	Whitish.	Insoluble.
Quinidia, $C_{20}H_{24}N_2O_2(HCl)_2PtCl_4$	26.8	Whitish. (Dry at 150° C.)	
Strychnia, $(C_{21}H_{22}N_2O_2.HCl)_2PtCl_4$	18.3	Yellow, crystallizable.	Insoluble.
Thebaina, $(C_{19}H_{21}NO_3.HCl)_2PtCl_4.H_2O$	18.7	Light yellow.	
Theobromina, $(C_7H_8N_4O_2.HCl)_2PtCl_4$	25.5	Brownish, floc. to cryst.	

143. Auric chloride gives precipitates in water solutions of salts of the greater number of the alkaloids, as follows. Many of the precipitates are soluble in alcohol. Some of them, on standing, separate the gold. The dried and ignited precipitates yield fixed quantities of metallic gold, according to the formulæ and percentages given:

	P. C. Au in pre.	Color, etc., of the pre.
Aconitia, $C_{30}H_{47}NO_7.HCl.AuCl_3$.	22.1	Light yel., reduced after a time.
Atropia, $C_{17}H_{23}NO_3.HCl.AuCl_3$.	31.3	Light yel.
Berberina, $C_{20}H_{17}NO_4.HCl.AuCl_3$.	29.1	Dark yel., insol. in **HCl.**
Brucia, ...	.	(Like Strychnia).
Caffeina, $C_8H_{10}N_4O_2.HCl.AuCl_3$.	37.0	Lem.-yel., cryst'e.
Cinchonia, ...	.	Yel., (like Quinia).
Cinchonidia, $C_{20}H_{24}N_2O(HCl)_2AuCl_3$	.	Yel., amorphous.
Codeina, no precipitate, . . .		(in concentrated solutions a brown precipitate).
Colchicin,		Slowly, yel. flocks; becom'g reduced.
Delphina, ...	.	Light yel.
Digitalin, ...	.	Slowly, a yellow cryst'e precip.
Emetia, ...	29.7	Light yel., amorp.
Hyoscyamia, ...	31.2	Yel.-white.
Morphia, ...	.	Light yel., dark'g, insoluble in cold **H Cl.**
Narceina, . . . - . . .	.	Yel., becom. red'd.
Papaverina, ...	.	Dark yel.
Physostigmia, ...	.	Red'ish-blue color, with reduction.
Quinia, ...	.	Light yel., amorp.

	P. C. Au in pre.	Color, etc., of the pre.
Quinidia, $C_{20}H_{24}N_2O_2(HCl)_2(AuCl_3)_2$	39.1	(Like Quinia) dry first in vacuo, then at 100° C.; melts at 115° C., or in boil. water.
Solania, no precipitate.		
Strychnia, $C_{21}H_{22}N_2O_2.HCl.AuCl_3$	29.2	Yel., amor., sol. in cold **H Cl**, slight. sol. in water, insol. in ether, sol. in alcohol, from which it cryst. orange.
Thebaina, ...		Red-brown.
Theobromina, ...		Slowly, slight, needle-form, cryst.
Veratria, $C_{32}H_{52}N_2O_8.HCl.AuCl_3$.	21.0	Clear yel., amorp.

GLUCOSIDES AND OTHER NEUTRAL BODIES: SOLID.

144. ABSINTHIN. $C_{16}H_{22}O_5$. A hard and obscurely crystalline solid of very bitter taste. Slightly soluble in water, very soluble in alcohol, soluble in ether, and soluble in aqueous alkalies. It is precipitated by tannic acid, not by subacetate of lead. When treated, dry, with concentrated **sulphuric acid**, and the mixture slightly diluted with water, a blue-violet color. It does not reduce potassio cupric sulphate, but reduces ammonio nitrate of silver to a mirror-coating.

145. ALOIN. $C_{17}H_{18}O_7$. A crystallizable, pale yellow solid, of neutral reaction and a taste at first sweet and then very bitter. It bears 100° C. without change. It is slightly soluble in cold water or alcohol, moderately soluble in the same when hot, and soluble (with a yellow color) in the **alkalies** and their carbonates.—Chlorine gas, in a solution of aloin, forms a bright yellow precipitate (chloraloil). Bromine also gives a yellow precipitate.—Concentrated **nitric acid** transforms aloin into chrysammic acid.

Chrysammic acid, $C_7H_2(NO_2)_2O_2$, is a yellow or greenish-yellow powder, of bitter taste and acid reaction, sparingly soluble in water, readily soluble in alcohol and in ether. It detonates when heated. Boiled with solution of **stannous chloride** it is precipitated as a deep violet powder. Chrysammate of **calcium** is a dark red insoluble powder.

146. AMYGDALIN. $C_{20}H_{27}NO_{11}$. A white, pulverulent, and crystalline solid, neutral, without odor, and with sweet and bitter taste. Soluble in 11 parts of water; sparingly soluble in cold, moderately soluble in hot alcohol; insoluble in ether.—Concentrated **sulphuric** acid colors it light violet-red. By boiling dilute sulphuric acid, it is transformed into oil of bitter almonds, glucose, and formic acid; by fermentation with emulsin, into bitter almond oil, hydrocyanic acid, and glucose. (16 parts of anhydrous amygdalin, as dried at 110° to 120° C., or 20 to 24 of ordinary commercial amygdalin, gives 1 part hydrocyanic acid and 8 parts of bitter almond oil.)—Permanganate of potassium forms cyanic and benzoic acids.

147. Asparagin. $C_4H_8N_2O_3(H_2O)$. Hard and brittle right rhombic (trimetric) crystals; inodorous and of slight taste. Soluble in 11 parts cold or 5 parts of boiling water (with slight acid reaction), insoluble in absolute alcohol, insoluble in ether, soluble in alkalies and acids. By fermentation with accompanying extractive substances, or with casein, succinate of ammonium is formed (sometimes with the intervening formation of aspartate of ammonium).

148. CANTHARIDIN. $C_5H_{12}O_2$. A colorless, odorless solid, crystallizing in rhombic tables or in needles, not volatile at 40° C., slightly volatile with water at 100° C., fusing and subliming at about 200° C. It acts as a vesicant on the skin. Insoluble in cold or warm water, sparingly soluble in alcohol, moderately soluble in ether, freely soluble in chloroform and benzole, soluble in oil of turpentine and in olive oil. Cantharidin has the relation of an acid of very weak power. Its potassium compound is soluble in 25 parts cold or 12 parts boiling water, in 3,300 parts cold or 110 parts boiling alcohol, insoluble in ether and chloroform. The barium cantharidate is insoluble in water and alcohol, as well as in ether and chloroform.—Cantharidin separations may be effected, first, by solution in aqueous potassa; then, after acidulating with sulphuric or phosphoric acid, by solution in chloroform.

149. CATHARTIC ACID (of senna-leaves). CATHARTIN.—An amorphous brown to black solid, soluble in aqueous alkalies and precipitated from this solution by acids. In its natural condition, partly combined with calcium and magnesium, it is soluble in water and insoluble in alcohol. Boiling dilute acids, in alcoholic solution, convert it—as a glucoside—into glucose and cathartogenic acid, a brown-yellow powder, insoluble in water, alcohol, and ether.

150. COLUMBIN. $C_{21}H_{22}O_7$. Colombo bitter.—A colorless solid, crystallizing in trimetric prisms, neutral, inodorous, and extremely bitter. It is sparingly soluble in cold water, alcohol and ether; more freely in aqueous potassa, being precipitated from the alkaline solution by addition of acids.—Strong **sulphuric acid** dissolves it with orange color, changing to deep red, and the addition of water to this solution causes a brown, flaky precipitate.

151. CUBEBIN. $C_{17}H_{16}O_5$. A white solid, crystallizing in small needles, melting at 120° C., inodorous, and tasteless. Slightly soluble in water and in cold alcohol, more soluble in boiling alcohol, soluble in 26 parts of ether and in acetic acid. It is precipitated from alcoholic solution by potassa. Concentrated **sulphuric** acid colors it bright red, soon changing to crimson.

152. ELATERIN. A colorless solid, crystallizing in hexagonal tables, fusible; insoluble in water, freely soluble in alcohol, sparingly soluble in ether. Precipitated from alcoholic solution by acetate of lead and nitrate of silver. Soluble in **sulphuric acid** as a red liquid, giving a brown precipitate on dilution with water.

153. FRAXIN. $(C_{16}H_{18}O_{10})_2H_2O$. A white solid, crystallizing in tufts of needles or right-rhombic prisms, of a slight acid reaction, inodorous, slightly bitter and astringent to the taste. It melts at 320° C., to a red liquid, solidifying amorphous, and dissolving in water with an orange color. At a higher heat it yields a crystalline sublimate, the aqueous solution of which, with ammonia, shows a yellow fluorescence. Sparingly soluble in cold, freely in hot water, moderately soluble in alcohol, slightly soluble in ether. The dilute aqueous solution has a blue or blue-green fluorescence, favored by alkalies but prevented by acids. The alcoholic solution is likewise fluorescent. It is turned yellow by fixed **alkalies** in aqueous solution, or by vapor of ammonia acting on the crystals; in aqueous solution **ferric** chloride causes a green color, followed by a yellow precipitate. Acetate of **lead** gives a yellow precipitate. Boiling dilute acids resolve it (as a glucoside) into fraxetin and glucose. Fraxetin, with strong nitric acid, shows successively dark violet, garnet-red, rose-red, and yellow colors, then becoming colorless.

154. LACTUCIN. A yellowish, fusible, bitter solid; crystallizable in rhombic plates; soluble in 80 parts of cold water, moderately soluble in alcohol, sparingly soluble in ether, soluble in acetic acid. Strong sulphuric acid turns it brown.

155. PHLORIDZIN. $C_{21}H_{24}O_{10}$. Crystallizes in silky needles or tufts, slightly bitter. Soluble in water, sparingly when cold but freely when hot; soluble in alcohol and methylic alcohol; insoluble in ether; soluble in aqueous alkalies. Dry phloridzin, treated with **ammonia** gas, on standing in the air, becomes, successively, orange, red, and blue (formation of phlorizein). Strong **sulphuric** acid colors it red. Dilute sulphuric acid, by prolonged contact, changes phloridzin to glucose and phloretin. The latter is an easily oxidizable substance, dissolving in ammonia, the solution soon depositing yellow scales.

156. POPULIN. $C_{20}H_{22}O_8$. A colorless solid, of a sweet taste, crystallizing in silky needles (with $2H_2O$), which become anhydrous at 100° C. and melt at 180° C. Populin dissolves in 2,000 parts of cold or 70 parts of boiling water, in about 100 parts of absolute alcohol, scarcely at all in ether, freely in moderately dilute acids, also in alkalies. It is precipitated from its acid solutions by alkalies, from its alkaline solutions by acids, and from its water solution by common salt. With concentrated **sulphuric acid** it forms a deep-red solution, from which water precipitates a red powder, soluble in water not acidulated but reprecipitated by acids. Fröhde's reagent colors populin violet. Boiling dilute mineral acids convert populin, as a glucoside, into benzoic acid, saligenin, and glucose. Boiling with hydrate of calcium resolves populin into benzoic acid and salicin.

157. QUASSIN. $C_{10}H_{12}O_3$. A colorless, inodorous, and very bitter solid, crystallizing in opaque, white, columnar prisms, melting when heated. It is soluble in about 200 parts of water of medium temperature; freely soluble in alcohol, slightly soluble in ether. Cold concentrated sulphuric acid dissolves it as a colorless liquid, from which water precipitates it unchanged.

Tannic acid precipitates it, both from aqueous and alcoholic solutions, but lead salts and mercuric chloride do not affect it. It does not reduce ammonio nitrate of silver.

158. SARSAPARILLIN. A colorless solid, crystallizable in needles; soluble in water and in alcohol, soluble in ether and in volatile oils. The solutions froth when shaken. Strong **sulphuric acid** dissolves it with deep red color, changing to violet and finally to yellow. From this solution water precipitates it unaltered.

159. TARAXACIN. Crystallizes in warty masses, of a pleasant bitter taste, fusible, and soluble in water, alcohol, ether, and in concentrated acids.

160. VANILLIN. $C_{10}H_6O_2$. Crystallizes in long, colorless, four-sided prisms, melting at 76° C. (or 82° C.), distilling with vapor of water, and subliming in part at 150° C. It is neutral in reaction, and has the characteristic odor of vanilla. It is nearly insoluble in cold, moderately soluble in hot water, freely soluble in alcohol, ether, and volatile oils. It dissolves in strong sulphuric acid and in potassa.

161. Separation of the Glucosides and Neutral Compounds—described in 144–160—by Water, Alcohol, Ether, and Aqueous Alkalies (recapitulation):

a. **Water** *dissolves* Absinthin (sparingly), Aloin (hot, sparingly), Amygdalin, Asparagin, Cathartin, Columbin (sparingly), Cubebin (slightly), Fraxin, Lactucin, Phloridzin, Populin (sparingly), Quassin (sparingly), Sarsaparillin, Taraxacin, Vanillin (slightly).—Water does *not* dissolve Cantharidin, Cubebin (except slight portions), Elaterin, Vanillin (except slight portions).

b. **Alcohol** dissolves Absinthin, Aloin, Amygdalin, Cantharidin (sparingly), Colombin (sparingly), Cubebin (sparingly), Elaterin, Fraxin, Lactucin, Phloridzin, Populin (sparingly), Quassin, Sarsaparillin, Taraxacin, Vanillin.—Alcohol does *not* dissolve Asparagin, Cathartin.

c. **Ether** dissolves Absinthin, Aloin (sparingly), Cantharidin, Cubebin, Elaterin (sparingly), Lactucin (sparingly), Sarsaparillin, Taraxacin, Vanillin.—Ether does *not* dissolve Amygdalin, Asparagin, Fraxin (except slight portions), Phloridzin, Populin, Quassin (except slight portions).

d. **Aqueous alkalies** dissolve Absinthin, Aloin, Asparagin, Cathartin, Colombin, Phloridzin, Populin, Vanillin; do *not* dissolve Cubebin.

[For Dragendorff's elaborate process for separation and identification of Foreign Bitters in Beer, see Arch. Pharm. (3), iii., 295; iv., 389; or Jour. Chem. Soc., 1874, p. 818; or Prescott's Exam. Alcoholic Liquors, N. Y., 1874.]

NITROGENOUS NEUTRAL BODIES.

162. ALBUMENOIDS. Varieties of Albumen, Fibrin, and Casein.—*Characterized* as infusible, non-volatile, amorphous solids, neutral in reaction and indifferent to combination; in natural condition, soluble in water containing alkalies or containing certain salts of alkalies; rendered insoluble in water by acids, and generally by heat in absence of dissolving agents, and by salts of certain heavy metals. Farther, they give a reaction for nitrogen (*a*), and color-tests with strong hydrochloric acid (*b*), and with mercuric nitrate in nitric acid (*c*), and act as reducing agents (*d*).—Albumenoids are distinguished and partly *separated* from Gelatin, Gums, and Dextrin, by their coagulation with heat or with mineral acids; they are separated from starch by dissolving in solution of alkali too dilute to affect the starch (see 176, *f*).

a. Mix the well-dried substance with dry **soda-lime,** introduce into a hard-glass long-necked matrass (or long and narrow test-tube), place a slip of red litmus-paper in the mouth, and heat, gradually, to incipient carbonization. Production of ammonia (this base being absent in the substance) indicates a nitrogenous organic body.—Albumenoids, on boiling with **potassa solution,** yield ammonia; a farther quantity being obtained by adding permanganate (Wanklyn).

b. Strongest hydrochloric acid dissolves albumenoids to a yellowish color, which becomes blue or violet by exposure to the air.

c. **Acid mercuric nitrate** solution—prepared by dissolving one part of mercury in two parts of nitric acid of spec. grav. 1.42 —on digestion with solid albumenoids, at 60° to 100° C. (140° to 212° F.), gives a deep red color. If the substance tested is in solution, it should be concentrated, and used in such small proportion that the reagent is not much diluted.

d. **Alkaline cupric solution** is turned violet by albumenoids, and on warming the cuprous oxide is quickly reduced. Solution of **permanganate** is also reduced by albumenoids.

Under the Microscope, albumenoids are turned yellow by iodine, and purple-violet by sulphuric acid with sugar.

163. Ovalbumen. Soluble in water with some turbidity and suspension of skinny particles; this solution being rendered nearly clear by alkalies or alkaline carbonates or common salt.—Chemically neutral water solutions are *coagulated* by heating to a very little above 63° C. (145° F.); by alcohol, carbolic acid, and creosote; by ether (but not completely), by nitric acid (quickly and completely), by hydrochloric acid (slowly redissolving when the acid is strong), and by sulphuric acid (slowly). Salts of silver, mercury, etc., coagulate it; also alum. Acetic and tribasic phosphoric acids do not coagulate it, but (by neutralizing the natural alkali) they render its pure water solution still more turbid. Tannic acid coagulates it quickly.—Strong **potassa,** or soda, gelatinizes albumen.

For weighing, albumen is precipitated from slightly acidulated solutions by boiling, washed with water, and dried first below 50° C. (122° F.), then at 100° C. (So treated, it is not rendered insoluble.)

164. Seralbumen. Dissolves in water with some turbidity.

Coagulated at 75° C. (167° F.), or, in presence of alkaline salts, at a higher temperature, while in presence of acetic acid a lower temperature suffices. Coagulated by dilute mineral acids, slowly or if heated quickly; redissolved by strong nitric acid and (readily) by hydrochloric acid. Coagulated by alcohol; not by ether (distinction from Ovalbumen).—Coagulated by salts of mercury and silver.—Aqueous alkalies dissolve coagulated seralbumen.

165. Casein. In natural condition, dissolved clear by water. *Coagulated* by rennet (separation from milk albumen); by moderately dilute acetic acid (separation from ovalbumen, seralbumen, milk albumen); by ether and by mineral acids and mercuric chloride. Not coagulated by dilute alcohol or by boiling (separation from seralbumen and from ovalbumen). Alkalies and strong acids, even strong acetic acid, dissolve coagulated casein. [Farther, see Phar. Jour., 1874, Sept. 5, p. 188.]

166. Milk Albumen. (0.3 to 0.5 per cent. of healthy cows' milk; found as high as 3 to 10 per cent. in diseased milk and in the colostrum.)—Not coagulated by rennet, but coagulated by boiling, after the slightest acidulation with acetic acid (two means of separation from Casein). Coagulated by mineral acids and salts of heavy metals; the coagulum being soluble in alkalies.

167. Determination of Casein and Albumen in Milk.—*a.* Take 50 grams of milk, add an equal quantity of water, add rennet, set aside at 40° to 50° C. Gather the precipitate (the casein with most of the fats) in a tared filter, wash with water, then with alcohol, then with ether thoroughly, dry at 110° C., and weigh as Casein.—To the filtrate from the curd (and first water washings), add 4 or 5 drops of acetic acid and boil. Gather the coagulum in a tared filter, wash with water, dry at 110° C., and weigh as Albumen.—(The filtrate from the curd of albumen is saved for determination of the Sugar, according to 187, *l*. This filtrate contains a minute proportion of an albumenoid called Lacto-protein, which is coagulated by mercuric nitrate—not by nitric acid.)

168. Quantitative Analysis of Milk.

(1) Determine the Total Solids, as directed in 64, *b*.

(2) Determine the Fats, as directed in 64, *b* (or *a*).

(3) Determine the Casein and Albumen, as directed in 167.

(4) Determine the Sugar, from the filtrate of 167, according to 187, *l*.

(5) Determine the Salts (soluble and insoluble in water). Evaporate 20 grams in a tared dish, with a tared small glass rod, ignite to whiteness (by triturating); weigh, then extract with water and dry and weigh the residue.

169. Commercial Examinations of Milk usually require, more especially, the following operations:

(1) Find the *volume per cent.* of cream. (Or use a lactoscope—64, *c*.)

(2) Take the *specific gravity*, and consider the relation between this and the amount of cream. Skimmed milk has a specific gravity about 0.004 greater than entire milk (from CHANDLER's averages.)

(3) For more exact data, find the *solids minus fats*, as directed in (1) and (2) of 168. The "solids not fat" is nearly the same proportion of the milk of different cows—also, of the whole milk, the skim milk, and the cream, alike (WANKLYN). Hence, variation in this quantity indicates sophistication.

(4) Examine with the microscope (presence of colostrum globules usually coinciding with excess of albumen), and test for impurities in general.

170. GELATIN. (Isinglass, Glue, "Gelatin.") An infusible, non-volatile, amorphous, horny solid; colorless to yellowish, translucent, brittle, odorless, and tasteless.—*Characterized* as a neutral and indifferent substance, evolving ammonia freely when heated dry with soda-lime or when boiled with potassa solution and permanganate (see Albumenoids, 162, *a*), and existing in a soluble and an insoluble condition.—Its soluble form *dissolves* very slowly and slightly in cold water, gradually and completely in boiling water, the solution if not very dilute congealing into a tremulous jelly-like mass when cold (distinction from Albumenoids which are coagulated by boiling).—Gelatin solution is *coagulated* by alcohol, by mercuric chloride, by chlorine gas, and especially by tannic acid (formation of leather) (the last, a *separation* from Gum arabic and Dextrin). In distinction, and separation *from Albumenoids*, it is not precipitated by nitric, hydrochloric, or sulphuric acids, or by salts of silver, copper, lead, iron, or aluminum. In not being precipitated by basic acetate of lead, it is separated from Dextrin, soluble starch, starch-paste, and Gums.—Gelatin solution dissolves the recent **cupric hydrate,** as formed in cupric sulphate solution by excess of potassa, turning the color to dark violet, which on warming becomes red, without precipitation of cuprous hydrate. It promptly reduces **permanganate** solution.

100 parts of gelatin, as dried at 130° C., by precipitation with tannic acid, yield about 135 parts of leather precipitate.

171. LEATHER yields its tannic acid to boiling dilute alcohol, the gelatin remaining coagulated. The dried and finely-rasped leather is first freed from oils and resins by digestion with ether free from alcohol, along with water. Untanned gelatin may be detected in leather, by the translucence of thin shavings (of the central portion), and by yielding a solution of gelatin when long macerated with water at about 90° C.

CARBOHYDRATES.

172. GUMS. Mostly $C_6H_{10}O_5$ (as an anhydride) in combination with alkaline-earthy bases or with water. *Characterized* as infusible and non-volatile, amorphous substances, destitute of nitrogen, more or less perfectly soluble in water, insoluble in absolute alcohol, ether, benzole, etc.; precipitated by subacetate of lead; and not readily transformed to glucose by boiling with dilute acids (a distinction from Dextrin and from Starch).

173. GUM ARABIC. Gum Acacia. Arabin, or Arabates of calcium, magnesium, potassium, etc.—Soluble (by digestion) in 2 parts of water, forming a syrupy liquid of spec. grav. 1.13, which mixes clear with $2\frac{1}{2}$ times its volume of 35 per cent. alcohol. Soluble in 20 to 25 parts of 45 per cent. alcohol. If acidulated (with mineral or acetic acids), arabic acid being liberated, it is much less soluble in dilute alcohol, *i.e.* requires for solution alcohol more dilute.

Gum arabic is *characterized* by a white precipitate by **subacetate of lead** or ammoniacal acetate of lead, in very dilute solutions; by giving (with oxalate) the reactions of *calcium* (distinction from Dextrin and Starch); by forming an almost insoluble jelly when in saturated solution it is treated with about $\frac{1}{12}$ volume of concentrated solution of **ferric chloride,** and by preventing the precipitation of iron salts by alkalies when in dilute solution (two points of distinction from Dextrin).

As a reducing agent, fresh solution of gum arabic, with potassio cupric solution, precipitates the cuprous hydrate after heating to 100° C. (Dextrin effecting this reduction at a gentle heat). Stale solution of gum arabic generally contains glucose.

Gum arabic gives no precipitate with tannic acid (*separation* from Gelatin and Ovalbumen); or with mineral acids (separation from Albumenoids); and no reaction with iodine (distinction from Starch).

Sulphuric acid, added to its one-half volume of concentrated acacia gum solution, turns it brown to black. Boiling with dilute sulphuric acid slowly transforms gum acacia (in part) to glucose.

Ordinary gum arabic, at 90° to 100° C., loses 10 to 15 per cent. of moisture; above 100° C., it is so altered as to be imperfectly soluble.

174. GUM TRAGACANTH is only in small part directly soluble in water, in which it swells to a jelly; the greater part dissolving by long boiling. The solution so formed consists chiefly of Arabin, with a very little Glucose, and gives reactions for these substances, according to their proportion. Boiling with dilute sulphuric or hydrochloric acid dissolves the gum more rapidly than with water, producing a little larger proportion of glucose. —The residue not soluble in pure water contains starch, and is colored blue by iodine.

175. DEXTRIN. British Gum. $C_6H_{10}O_5$.—A yellow-white to colorless amorphous solid; tasteless and odorless. It is *soluble* in about one part of water, to a syrupy semi-liquid, which is miscible with 1½ volumes of 60 per cent. alcohol or with 3 volumes of 50 per cent. alcohol. It is insoluble in 90 per cent. alcohol, sufficient of which precipitates it from solutions not too dilute; and insoluble in ether, chloroform, bisulphide of carbon, etc.—Commercial dextrin almost always contains glucose; frequently contains "soluble starch" (15 per cent. of which is held not objectionable); and is sometimes brown from presence of caramel.

Concentrated **sulphuric acid** dissolves dry dextrin, without color in the cold but with blackening when warmed.—Subacetate or ammoniacal **acetate of lead** precipitates dextrin from very dilute solutions (in cold and dilute solution, a distinction from Glucose).—Pure dextrin (free from glucose) reduces **potassio cupric sulphate** at 80° to 90° C. It does not reduce boiling solution of cupric acetate (distinction from Glucose).—Pure dextrin is not colored by **iodine** (distinction from Starch and "soluble starch"); nor precipitated by **tannic acid** (*separation* from Starch and soluble starch, Gelatin, and Ovalbumen); nor by mineral acids (separation from Albumenoids); nor by baryta water (separation from Soluble Starch).

Dextrin is dried (over a glycerin-bath) at 110° C. Its precipitate by subacetate of lead is $\mathbf{Pb\ C_6H_{10}O_6}$.

176. STARCH. Chiefly $C_6H_{10}O_5$; being an organized body, of many varieties of structure, and containing cellulose in the envelopes of the granules.—Varieties of starch are identified by their form under the microscope (*a*). Starch in general is *characterized* by its relations to solvents (*b*); its color with iodine (*c*); its precipitates with tannic acid, subacetate of lead, and baryta (*d*); and its easy transformation to "soluble starch," dextrin, and then glucose (*e*).—Starch-paste and "soluble starch," both, are *distinguished and in part separated* from Albumenoids by non-precipitation with heat, or with mineral acids (*e*); from Gelatin by precipitation with subacetate of lead (*d*); from Gums by precipitation with tannic acid, and from Dextrin by precipitation with tannic acid or with baryta water (*d*). The complete *separation* of starch from Albumen, Gelatin, or Gum is effected by first changing it to glucose (*e*) and then washing the latter away (from the coagulum) with strong alcohol.—Starch is separated from Grains or other parts of Plants by water-washing (*f*), and *determined* directly or as glucose (*g*).

a. The *starch granules* are from $\frac{1}{3000}$ to $\frac{1}{260}$ inch in diameter, flattened and ovate, with concentric rings (the borders of overlapping layers), and mostly with a small eccentric nucleus. They are characteristic of each variety.

b. Natural starch is insoluble in water, alcohol, ether, etc. **Water** at 60° to 75° C. (140° to 167° F.) bursts the granules of natural starch; a small part of which is apparently dissolved, the larger part remaining suspended in minute particles forming a gelatinous semi-solution, while a small portion, consisting of the envelopes, readily subsides, the whole being known as *Starch-paste.* Boiling water slowly changes starch-paste to "soluble starch" and to Dextrin.—Caustic **potassa** solution of 2 or 3 per cent. causes starch to swell to starch-paste; finally forming some "soluble starch."—When starch is triturated with two-thirds its weight of **concentrated sulphuric acid,** in the cold, and left for an hour, then washed on a filter with alcohol till free from acid, it is transformed into "*Soluble Starch.*"

This is a modification of starch, soluble in cold or hot water to a syrupy liquid not quite so clear as dextrin; colored blue to violet with iodine (distinction from Dextrin); precipitated by alcohol when the latter is as much as 50 per cent. (dextrin requires stronger alcohol for precipitation); precipitated by tannic acid and by baryta water (two ways of separating from dextrin); precipitated by subacetate of lead (coinciding with dextrin).

Concerning solution of starch by its transformation into Dextrin and Glucose, see *e*.

c. Free **iodine**—in solution with water or alcohol or water with iodide, or in vapor—colors starch blue to violet, forming the "iodide of starch" (a product of adhesion). The color is destroyed by heating (returning when cold), by washing with alcohol, and by chlorine, potassa, hydrosulphuric acid, or other agents which bring the iodine into chemical combination.

d. **Tannic acid** precipitates starch-paste; the precipitate being soluble in excess of the starch, and soluble by heat—separating again when cold. **Baryta** water, and solution of **subacetate of lead** or ammoniacal solution of acetate of lead, precipitate starch-paste (as well as soluble starch).

e. Starch is *changed to Glucose* (through soluble starch and dextrin) very quickly by boiling **dilute mineral acids** (two to three per cent.); very slowly by boiling with water, and quite effectually by the conditions of the alcoholic and "saccharine" fermentations.

f. *Cereal grains*, or other parts of plants, are finely pulverized, and then washed on a hair sieve with cold water, and the washings allowed to subside (as in manufacture). The starch residue may be washed again through a bag of fine linen. The residue is then washed on a filter with 45 per cent. alcohol containing 0.1 per cent. potassa, then with 60 per cent. alcohol, then with ether; and dried, first below 60° C., lastly at 100° to 110° C., when it may be weighed, as starch.

g. Starch may be determined as Glucose (187, *l*); after boiling with dilute sulphuric acid (*e*) and neutralizing. $C_6H_{12}O_6$: $C_6H_{10}O_5$: : 180 : 162.

177. PECTOUS SUBSTANCES. Vegetable products corresponding in properties to the gelatinoids of the animal kingdom.

178. PECTOSE. Insoluble in water, alcohol, or ether. Dissolved as Pectin, etc., by long boiling with **water,** more readily with vegetable acids. Hot dilute mineral **acids** dissolve pectose as Pectin, which by longer treatment becomes Metapectin. **Alkalies,** by hot aqueous digestion, form soluble salts of Metapectic acid.

179. PECTIN. Neutral; soluble in cold or hot **water;** gelatinized by dilute alcohol and precipitated by strong alcohol; changed by hot mineral **acids** to Metapectic acid; changed by cold dilute **alkalies** into soluble salts of Pectic acid, by hot and strong alkalies into soluble salts of Metapectic acid.

180. PECTIC ACID. In its moist state, gelatinous. Neutral in reaction. Insoluble in cold and scarcely soluble in hot **water;** by boiling water slowly changed to soluble Parapectic acid, afterward to Metapectic acid. Pectic acid jelly is hardened and parapectic acid solution is precipitated by alcohol and by solution of **sugar.** Boiling with dilute **acids** readily converts pectic acid to Metapectic acid. **Alkalies,** on contact with pectic acid, form pectates soluble in water but insoluble in alcohol. The pectates of non-alkaline metals are insoluble in water. Boiling with aqueous alkalies converts pectic acid into soluble salts of Metapectic acid.

181. PARAPECTIN is neutral, soluble in **water,** insoluble in alcohol, by which its aqueous solution is gelatinized. Boiling dilute **acids** convert parapectin into Metapectin. Aqueous **alkalies,** on contact with parapectin, form soluble salts of Pectic acid.

182. PARAPECTIC ACID is soluble in **water** (with acid reaction), the solution changing into one of Metapectic acid. Parapectic acid is precipitated from water solution by strong alcohol. It forms soluble salts with the **alkalies;** insoluble salts with the other metallic bases.

183. METAPECTIN is soluble in **water** (with acid reaction), insoluble in alcohol. **Alkalies** form with it the soluble salts of Pectic acid.

184. METAPECTIC ACID is producible from all pectous substances, but produces none of them. It is soluble in **water** (with acid reaction); soluble in **alcohol** (separation from all other pectous substances); and forms soluble normal salts with all the **bases** (the non-alkaline salts of other pectous acids being insoluble.)

Solution of **subacetate of lead** precipitates all the pectous substances (including metapectic acid). Hot **potassio cupric solution** is reduced by all the pectous substances. They are but slightly or not at all changed to Glucose, by boiling dilute acids.

185. CELLULOSE. $(C_6H_{10}O_5)n$. *Characterized* by its physical properties and relations to solvents (*a*); by its transformation into parchment-paper (*b*), and into dextrin and glucose (*c*), and by its formation of gun-cotton (*d*). It is *separated* from Starch by its solubility in ammonio cupric solution (*a*), and by its insolubility in hot dilute acids.

a. Pure cellulose is a white, translucent solid; of specific gravity about 1.5; insoluble in water, alcohol, ether, oils, and other neutral solvents. It is slowly disintegrated and partly dissolved with decomposition by strong aqueous alkalies. Hot dilute mineral acids scarcely affect it; moderately dilute nitric acid changing it to Xyloidin.—Finely divided cellulose slowly dissolves in a solution of oxide of copper in strong ammonia; being precipitated therefrom unchanged by hydrochloric acid.—Fibres of cellulose, superficially softened by sulphuric acid, or by potassa solution, are colored violet to blue by iodine solution, and are by this means rendered distinctly visible under the microscope. Also, by dipping in a 1 per cent. solution of potassium iodide and drying, then immersing in strong sulphuric acid and washing with water, cellulose is converted into a blue substance, showing red and blue globules under the microscope (TERRELL).

b. Sulphuric acid of about 1.5 or 1.6 spec. grav., acting for a very short time on cellulose (unsized paper), changes its state of aggregation so as to form parchment-paper.

c. Concentrated sulphuric acid, in the cold, slowly dissolves (thoroughly dry) cellulose to a colorless syrup, which closely resembles dextrin. It is, however, colored blue, or after standing some days in the acid, violet to brown, by iodine. The name *amyloid* has been applied to this substance. If it is now, after several days' contact of the acid, diluted with 30 or 40 parts of water and boiled (until a portion is not precipitated by strong alcohol), it is wholly converted into *glucose*.

d. Nitric acid of spec. grav. 1.5, or a mixture of nitrate of potassa 2 parts and concentrated sulphuric acid 3 parts, at a temperature below 50° C. (122° F.), converts clean, dry cotton wool (finely divided cellulose), by 24 hours' contact, into *nitrocellulose*. This is washed first with cold water, then with hot water, lastly with alcohol and dried at ordinary temperature.

186. Nitrocellulose, Pyroxylon, or Gun Cotton is the substitution of $(NO_2)_{7-9}$ for H_{9-7} in $C_{18}H_{30}O_{15}$—, the lower substitutions being most soluble in ether, the higher substitutions being most explosive. It is more readily soluble in alcoholic than in pure ether—formation of Collodion. It is not attacked by dilute acids or alkalies : strong sulphuric acid dissolves it slowly, strong alkalies dissolve it with decomposition.—The residue from collodion is unchanged pyroxylon, in a firm and elastic mass, capable of being moulded at about 140° C.

187. GLUCOSE. $C_6H_{12}O_6.H_2O$. Grape sugar. Starch sugar. Dextrose.—*Characterized* by its physical properties and solubilities (*a*) ; its rotation of polarized light (*b*) ; its reactions with potassa (*c*) and, *as a reducing agent*, with potassio cupric solutions (*d*), cupric acetate (*e*), ferricyanide of potassium (*f*), ammonio silver nitrate (*g*), bismuthic subnitrate (*h*), and molybdate of ammonium (*i*). It precipitates ammoniacal acetate of lead (*j*), and reacts with stannic chloride and cobaltous hydrate (*k*).—*From Sucrose*, it is *distinguished* by a stronger reducing power (*d, e, f, g, i*), by not blackening with concentrated sulphuric acid (189, *c*), but turning brown with potassa solution (*c*).—*From Lactose*, it is distinguished by stronger reducing power (*e, i*), less soluble precipitate with ammoniacal acetate of lead (*j*), and by not blackening with concentrated sulphuric acid.—*From Fructose*, it is *separated* by crystallization, and *distinguished* by contrary rotation (*b*).—It is *separated* from Dextrin, Soluble Starch, Gums, the Pectous substances save metapectic acid, Gelatin, and Albumenoids, by solution in 90 per cent. alcohol (*a*); from Fats, etc., by insolubility in ether.—It is *determined* by the volumetric solution of potassio cupric salt (*l*), or by the polariscope (*b*), or by fermentation (*m*).

a. Glucose *crystallizes*, with some difficulty, in warty or cauliflower-like masses, hydrated; but from strong alcohol, in anhydrous needles. At 60° C., the hydrate becomes an anhydrous, white powder; at 100° C., the hydrate *melts* to a transparent mass; but the anhydrous glucose melts at 130° C. For *weighing*, it should be well dried at 60° C., then at 110° C. (without melting).—Glucose is *soluble* in a little more than one part of cold water; a saturated solution having a spec. grav. 1.206 and containing 45 per cent. of anhydrous glucose. Dilute alcohol dissolves it freely; 100 parts of 90 per cent. alcohol dissolve 2 parts in the cold, 20 parts with boiling; in cold, absolute alcohol it is scarcely at all soluble. Insoluble in ether, chloroform, oils; soluble in 60 parts hot amylic alcohol; soluble in methylic alcohol.

b. Anhydrous glucose has a specific rotatory power of 55° (Pasteur) to the right.

c. **Potassa**, or milk of lime, when warmed in solution of glucose, causes a reddish-yellow to brown color with deposition of a humus-like substance (distinction from Sucrose).

d. The test for *reduction of cupric hydrate* to cuprous hydrate in presence of alkali may be made by adding a drop or two of cupric sulphate solution and then an excess of potassa, or by use of enough of the standard solution specified in *k* to tinge the test-liquid bluish. At a gentle heat (short of boiling) glucose throws down the brownish-yellow precipitate of cuprous hydrate, changed by boiling to a brownish-red precipitate of cuprous oxide. Without heat, the reduction occurs after standing some time. (Compare Sucrose, *b*.)

e. Solution of **cupric acetate** is reduced by glucose on boiling (distinction from Sucrose and from Lactose—the latter effecting a slight reduction after long boiling).

f. **Ferricyanide of potassium** (1 part) in solution with potassa (½ part), at 80° to 100° C., is reduced by glucose to ferrocyanide. The reduction is shown by loss of color, and by a blue precipitate with ferric salt. (Distinction from Sucrose and from Dextrin.)

g. Boiling solution of glucose separates silver (black) from **nitrate of silver**; more readily blackens the recent oxide of silver, and gives a dirty gray precipitate in solution of ammonio nitrate of silver (the latter a means of distinction from Sucrose).

h. Basic **bismuthic nitrate**, with carbonate of sodium, is reduced by boiling solution of glucose, with precipitation of bismuthous oxide as a dark gray sediment.

i. Solution of **molybdate of ammonium,** at boiling heat, is reduced by glucose, with formation of the blue molybdic molybdate (distinction from Sucrose, Lactose, and Dextrin).

j. Ammoniacal acetate of **lead** solution is precipitated by addition of concentrated solutions of glucose, the precipitate dissolving in excess of glucose solution, but appearing again on boiling in solutions not too dilute and remaining when cold.

k. **Stannic chloride** blackens when warmed with glucose.—**Nitrate of cobalt** in concentrated solution of glucose is not colored by addition of solid potassa and boiling (with pure Sucrose a violet-blue precipitate is obtained).

Quantitative.—*l*. Glucose is determined in its reduction of copper by use of a *standard solution* made as follows: 34.64 grams pure crystallized cupric sulphate dissolved in 200 c.c.

water, with 150 grams ~~neutral potassic tartrate~~ in about 500 c.c. of a 10 per cent. solution of soda (sp. gr. 1.14), the mixture diluted to 1 litre. 1 c.c. is reduced by 0.005 gram of (anhydrous) glucose, or by 0.0067 gram of lactose.* The solution must not suffer change by boiling. The addition of about 100 c.c. of pure glycerin (in the litre) prevents decomposition.

The solution of sugar is diluted to such a number of times its own volume that it shall not be far from 1 per cent. glucose. Then, 10 c.c. of the blue solution are taken in an evaporating-dish, 40 or 50 c.c. of water added, and, while boiling, the graded sugar solution is added, until no blue color remains (after the precipitate has subsided or been filtered out). The quantity of sugar solution used contains 0.05 grams glucose, or 0.067 grams lactose.

m. Pure sugar may be determined by *fermentation*, in a Will's Fresenius' carbonic acid apparatus, as follows: In the first flask, of about 60 c.c. capacity, place 33.3 grams of the solution to be determined, and which is made of 5 to 10 per cent. strength of sugar. Add 0.3 gram tartaric acid and a small pinch of good pressed yeast, close the first flask (so that gas must pass through sulphuric acid in the second flask), and weigh the apparatus. Set aside at 30° to 35° C. (86° to 95° F.) for three days; and weigh again. The weight of carbonic anhydride lost, multiplied with 2.0454, gives the amount of anhydrous glucose, or of crystallized lactose, and, if multiplied by 1.9432, the quantity of sucrose. The results are not close.

188. LACTOSE. $C_6H_{12}O_6$ (crystallized). Milk Sugar.—*Characterized* by its physical properties (*a*); its reactions as a reducing agent (*b*), and with acids and alkalies (*c*); with ammoniacal acetate of lead and with lime (*d*); and by its fermenta-

* That is, 180 parts of glucose ($C_6H_{12}O_6$), or 240 parts of lactose ($\frac{4}{3}$ of $C_6H_{12}O_6$), suffice to consume 40 parts of oxygen ($2\frac{1}{2}O$), reducing 1247 parts ($5\,CuSO_4[H_2O]_5$) of copper salt. And 180 : 1247 : : 5 : 34.64.

tions.—It is *distinguished* from Glucose by a somewhat weaker reducing power (*b*), a more sparing solubility in cold water or dilute alcohol (*a*), and by blackening with sulphuric acid (*c*); from Sucrose by greater reducing power (*b*) and insolubility in strong alcohol. It is *determined* volumetrically by the potassio-cupric solution (see Glucose, *l*).

a. Lactose crystallizes in hemihedral trimetric crystals, hard and colorless, becoming anhydrous ($C_{12}H_{22}O_{11}$) at 150° C., and turning brown without melting at 160° C.—It is *soluble* in 6 parts of water at ordinary temperature or 2½ parts hot water, the cold saturated solution having a maximum spec. grav. 1.060, and is insoluble in cold absolute alcohol and in ether.

b. The **potassio-cupric** solution is *reduced* by lactose very nearly as readily as by Glucose (187, *d* and *l*) (distinction from Sucrose); one-third greater quantity being required, however, to produce the same effect.—Solution of **cupric acetate** is only reduced very slightly and slowly by boiling with lactose (distinction from Glucose).—**Molybdate of ammonium** solution is scarcely changed in a perceptible degree by boiling with lactose (distinction from Glucose).—Ammoniacal **nitrate of silver** solution is reduced by boiling with lactose (distinction from Sucrose).

c. Concentrated **sulphuric acid** blackens lactose, rapidly when warmed (distinction from Glucose).—Potassa slowly turns lactose solution brown after heating to boiling point (distinction from Glucose).

d. **Ammoniacal acetate of lead** solution gives but a slight precipitate, soluble in water and not reprecipitated on boiling. With milk of **lime,** not in excess, lactose forms a compound soluble in water, insoluble in alcohol.

189. SUCROSE. $C_{12}H_{22}O_{11}$. Cane Sugar. Saccharose.—*Characterized* by its physical properties (*a*); its reactions as a reducing agent (*b*); its reactions with alkalies and acids (*c*), and with ammoniacal acetate of lead (*d*). *From Glucose* it is *distinguished* as a less powerful reducing agent (*b*), by blackening with sulphuric acid or turning brown with potassa solution (*c*), and by its reaction with cobalt (*e*). It is distinguished from Lactose by weaker reducing power (*b*). It is approximately *separated* from Lactose by solution in cold water, and fully separated from Dextrin, Gums, Gelatin, and Albumenoids by solution in 90 per cent. alcohol. It is separated from Fats, Resins, etc., by not dissolving in (nearly absolute) ether. It is *determined* by volumetric solution of potassio copper salt, after being changed to glucose (*c*, and 187, *l*), by the specific gravity of its pure water solutions, by its specific rotatory power as measured in the polariscope, and by fermentation as directed for Glucose, 187, *m*.

a. Sucrose *crystallizes* readily in monoclinic (rhomboidal) prisms, generally with hemihedral faces, and anhydrous. At 160° C. (320° F.) it *melts* to a clear liquid which solidifies to "barley sugar"; at about 210° C. (410° F.) *Caramel* and other products are formed.—Sucrose is *soluble* in about $\frac{1}{3}$ part of water; scarcely soluble in cold absolute alcohol, insoluble in ether, chloroform, benzole, etc.—Sucrose has a specific rotatory power of 73.8° to the right.

b. **Potassio cupric** solution is at first not at all reduced by sucrose on warming, or even on digestion over the water-bath, but after boiling 5 or 10 minutes, a slight precipitate of cuprous hydrate appears, (distinction from Glucose, Lactose, and Dextrin). —Solution of **acetate of copper** is not reduced by long boiling (distinction from Glucose).—**Ferricyanide of potassium** is not reduced to ferrocyanide by hot solution of sucrose (distinction from Glucose).—**Stannic chloride** is reduced on warming, and **chromate** with excess of potassa on boiling, with sucrose, (reactions coinciding with those of Glucose and Lactose).—Ammoniacal **nitrate of silver** solution is not reduced, though turned yellowish, on warming with sucrose (a distinction from Glucose). Recent oxide of silver with excess of potassa is blackened on boiling with sucrose.—**Molybdate** of ammonium (neutral solution) is unchanged by sucrose (distinction from Glucose).

c. Sucrose is not readily colored by warming with solution of **potassa** (distinction from Glucose). **Lime** forms a soluble compound with sucrose.—**Concentrated sulphuric acid** blackens sucrose on warming, with separation of carbon and evolution of sulphurous and formic acids (distinction from Glucose).—**Dilute minreal acids** (2 to 3 per cent.), boiled 10 to 15 minutes with sucrose, *transform it into glucose.* The same change is very slowly effected by long boiling in water, and with moderate rapidity by boiling with dilute vegetable acids. Also by the conditions of alcoholic fermentation.

d. **Ammoniacal solution of acetate of lead** gives a white precipitate ($Pb_2C_{12}H_{18}O_{11}$), scarcely soluble in cold but readily soluble in hot water.

e. The blue to violet and rose-red precipitate made by adding potassa to **nitrate of cobalt** solution and boiling is scarcely altered by presence of sucrose, or held a little more in the violet. (In presence of Glucose, the mixture after boiling is colorless or brownish, but not violet or blue.)

CARAMEL. A mixture of three compounds:

Caramelane—brittle at ordinary temperatures, soft at 100° C., odorless and bitter; deliquescent and very soluble in water, sparingly soluble in alcohol, insoluble in ether.

Caramelene—brittle, freely soluble in water, not deliquescent, sparingly soluble in alcohol, insoluble in ether.

Caramelin—black, shining, and infusible; having three modifications with different and varying solubilities.

Caramel is precipitated by subacetate of lead solution; and reduces potassio cupric solution. As generally prepared, caramel has a characteristic, "burned-sugar" odor.

190. MANNITE. $C_6H_{14}O_6$. Crystallizes readily from solution in thin, four-sided prisms; melts at 160° C., and at 200° C. (392° F.) distils with little decomposition. It dissolves in 6 or 8 parts of water of ordinary temperature, in 80 parts of 60 per cent. alcohol or 1400 parts of absolute alcohol or smaller quantities of boiling alcohol, but is insoluble in ether.—It is not blackened by concentrated sulphuric acid, or turned brown by boiling with potassa, and it does not reduce the potassio cupric sulphate solution. It is not subject to the alcoholic fermentation.

ALCOHOLS AND THEIR PRODUCTS.

191. METHYLIC ALCOHOL. CH_4O. *Recognized* by its sensible and physical properties (*a*); its reaction with potassa and, as a commercial article, with sulphuric acid (*b*); by solution of recent mercuric oxide (*c*); by its reducing power (*d*), and its formation of formic acid (*e*). It is *separated* by fractional distillation (*f*). It is approximately *determined* as methyl oxalate (*g*) or as formic acid (*e*, and Formic acid *j* or *k*).

a. Pure methylic alcohol is a colorless *liquid*, of spec. grav. 0.800, boiling at 66° C. (151° F.), and of characteristic taste and odor. The commercial article is seldom free from empyreuma. It is miscible in all proportions of water, alcohol, and ether, and dissolves resins and nearly all substances soluble in ethylic alcohol.

b. The addition of **potassa**, with boiling by the heat of the water-bath, causes a brown color in a short time (Ethylic alcohol only after a long time).—Ordinary methylic alcohol gives a red to red-brown color with concentrated **sulphuric acid**.

c. Add (to the distillate *f*) 2 or 3 drops of very dilute solution of **mercuric** chloride, then solution of potassa in excess, agitate and warm. If methylic alcohol is present, the mercuric oxide will be dissolved.

d. Methylic alcohol readily decolorizes **permanganate** of potassium solution; but does not reduce silver nitrate, or potassio cupric solution.

e. Oxidation to *formic acid* is effected by distillation of 2 c.c. of the liquid examined, in a retort of 60 c.c. capacity, with 2 grams of powdered bichromate, 15 c.c. of water, and 25 drops of sulphuric acid—digesting fifteen minutes and then distilling 15 c.c.

f. In the *distillation* of methylic alcohol, add a little animal charcoal and a little solution of sodic carbonate, and receive the distillate at 66° to 76° C. (151° to 169° F.)

Quantitative.—*g.* Place in a retort 55 grams crystallized oxalic acid and the mixture of 35 grams of concentrated sulphuric acid and 25 grams of distillate *f*, digest for ten hours, and distil from an oil-bath at 160° to 180° C., as long as anything passes over. The distillate consists of oxalic ethers; methyl oxalate being freely soluble in water, while ethyl oxalate is nearly insoluble. The distillate is now washed with 25 times its volume of water; the clear solution decanted, digested, in a close bottle, with excess of potassa, the mixture acidulated with acetic acid and precipitated with calcium chloride (adding potassic acetate). Gather the oxalate of calcium, wash, dry, and ignite to carbonate (adding ammonium carbonate and igniting slightly again, if necessary). $CaCO_3 : 2CH_4O :: 1 : 0.64$.

192. ETHYLIC ALCOHOL. C_2H_6O. *Characterized* by its physical and sensible properties (*a*); by the extent of its reducing power (*b*); by its formation of iodoform (*c*); of various compound ethers (*d*), and of acetic acid (*e*).—*Separated* by fractional distillation, solubility in water, and insolubility in fixed oils. Separated from methylic alcohol as an oxalic ether (191, *g*), from amylic alcohol by solution in water or by fractional distillation.—*Determined* by the specific gravity or by the boiling point of its mixtures with water.

a. A transparent, limpid liquid, of spec. grav. 0.794, freezing at -95° C. and boiling at 78° C. (173° F.), of an agreeable and pungent odor and a sharp and burning taste. It is miscible with water, ether, chloroform, benzole, petroleum naphtha, volatile oils and castor oil, and dissolves resins and camphors.

b. Alcohol—as a hot liquid or as vapor—slowly reduces **chromic acid,** or a mixture of potassic bichromate and sulphuric acid—the alcohol being first oxidized to acetic acid. (This is in common with aldehyde, acetic acid, formic acid, and many volatile organic bodies.) **Permanganate** of potassium is but slowly reduced by ethylic alcohol—so that the red tinge of a slight addition of a $\frac{1}{1000}$ solution is scarcely at all affected for several minutes. (Methylic alcohol, Formic acid, Aldehyde, and many other volatile organic bodies, more readily reduce the permanganate.)

c. The production of iodoform from alcohol is a result (in part) of the reducing power of the latter upon alkaline iodate:

$$6KHO+6I=5KI+KIO_3+3H_2O$$
$$KIO_3+C_2H_6O+2I=CHI_3+KCHO_2+2H_2O$$

Take 3 to 5 c.c. of the distillate to be tested, 5 to 6 drops of a 10 per cent. potassa solution; warm to 100° or 120° C. (212° to 248° F.), and add—of a solution of potassic iodide in five parts of water, saturated with iodine—until the liquid is brownish-yellow. If, on agitation, the color does not disappear, add one or two drops of the potassa solution. If alcohol is present, the iodoform appears, sooner or later, in yellow scaly particles. With a power of 200 to 400 diameters, these are seen as hexagonal stars and rosettes. Iodoform is formed also by Aceton, Aldehyde, Acetic ether, Butyric alcohol, Amylene. Not formed by Ether, Amylic Alcohol, Chloroform, Chloral, Chloral Hydrate, and, according to Lieben, not formed by Methylic alcohol.

d. See under Acetic acid, 40, *b*, and Butyric acid, 41, *b*. (One c.c. of the distillate to be tested is treated with 0.3 to 0.5 gram of dry potassic acetate and 2 or 3 c.c. of sulphuric acid.)

e. Acetic acid is formed from alcohol by digestion with a mixture of **bichromate** of potassium and dilute **sulphuric** acid, or of **permanganate** of potassium and dilute sulphuric acid. See 40.

193. ALDEHYDE. C_2H_4O. Acetic Aldehyde.—A transparent and colorless liquid, of spec. grav. 0.800 at 0° C., *distilling* at about 21° C. (70° F.), neutral in reaction, of a pungent and suffocating odor, slightly resembling that of apples. The vapor irritates the eyes.—It is miscible in all proportions with water, alcohol, and ether, but not with aqueous chloride of calcium (separation from Alcohol). It dissolves sulphur, phosphorus, and iodine.—It promptly *reduces* ammonio nitrate of **silver**, forming a specular coating on the glass (distinction from Acetic acid, Alcohol, Ether). It burns readily, with a blue flame. It is blackened by **sulphuric** acid.—**Potassa** solution, warmed with aldehyde, colors it brown, with deposition of "aldehyde resin" and formation of acetate and formate (a characteristic test). Ammonia (gas) with aldehyde forms aldehyldate of ammonium, a compound of an ammoniacal, terebinthinate odor, crystallizing (from ether or alcoholic ether) in transparent acute rhombohedrons, melting between 70° and 80° C., and distilling at 100° C. It dissolves in water, sparingly in alcohol and ether. With other bases aldehyde acts as a monobasic acid, exchanging one atom of its hydrogen.

194. SULPHETHYLATES. $HC_2H_5SO_4$. Ethyl-sulphates.—Sulphethylic acid is a limpid, oily, acid *liquid*, of spec. grav. 1.315, decomposed by heat, evolving ether at 130° to 140° C. (266° to 284° F.)—It is *soluble* in water and alcohol, not in ether.—Its metallic *salts* are all soluble in water, and are mostly soluble in aqueous but not in absolute alcohol, the ammonium salt only is soluble in ether. The sulphethylates are gradually decomposed in boiling water. *Barium* sulphethylate crystallizes in permanent monoclinic prisms, with $2H_2O$ which is expelled in a vacuum, the anhydrous salt bearing 100° C. without change. It dissolves in about one part of water, and in a larger quantity of aqueous alcohol. The *sodium* salt crystallizes in slightly efflorescent hexagonal plates, with H_2O, soluble in less than one part of water, melting at 86° C. When anhydrous, it bears 100° C. without change.

195. ETHER. $(C_2H_5)_2O$. *Recognized* by its sensible and physical properties (*a*). *Separated* by distillation, or by solution (*b*).

a. Spec. grav., at 15° C., 0.713; at 17.5° C., 0.7185. Boiling point, 35° C. (95° F.) Ether of spec. grav. 0.728, and boiling at blood heat, has from 5 to 6 per cent. of about 90 per cent. alcohol; that of spec. grav. 0.750 has about 25 per cent. of 88 per cent. alcohol.—At 17.5° C. (63.5° F.), one part of ether dissolves in 12 parts of water, and 35 parts of ether dissolve one part of water. Alcoholic ether is more soluble in water:

"Ether"	of	sp. gr.	0.719	to	0.721	dissolves	in	12.0	parts water.
"	"	"	0.724	"	0.726	"	"	10.0	" "
"	"	"	0.729	"	0.731	"	"	7.7	" "
"	"	"	0.733	"	0.735	"	"	6.2	" "
"	"	"	0.738	"	0.741	"	"	5.0	" "
"	"	"	0.743	"	0.746	"	"	4.3	" "
"	"	"	0.748	"	0.750	"	"	3.8	" "

Salts not soluble in ether (as dry potassic carbonate) separate it from water almost wholly. Ether is miscible in all proportions with alcohol, chloroform, benzole, petroleum naphtha, fixed and volatile oils, and dissolves resins, sulphur, phosphorus, iodine, and ferric, mercuric, and auric chlorides. **Tannic acid** does not dissolve or deliquesce in absolute ether, but deliquesces in the "stronger ether" of spec. grav. 0.728. Ether is less soluble in glycerin than in water. It mixes with concentrated sulphuric acid, the liquid turning brown when warmed.—In the air, ether very slowly oxidizes to acetic acid. Its combustibility renders it necessary to use strict precautions in the manipulation of its vapor.

b. Ether is approximately separated from alcohol by means of glycerin (or water). A test-tube of over 20 c.c. capacity is graduated from the point of 10 c.c. contents (marked 0) to the point of 20 c.c. contents (marked 10). Ten c.c. of glycerin or water is taken in the tube, then 10 c.c. of the ether is added, the contents shaken together, the ether allowed to separate, and the increase in the lower layer is read off.

196. NITROUS ETHER. $C_2H_5NO_2$. Nitrite of ethyl.—*Characterized* by its sensible and physical properties (*a*) and by reactions of nitrites.—*Estimated*, in its alcoholic mixtures, by their boiling point (*b*), and by volumetric trial with permanganate (*c*).

a. Nitrite of ethyl is a yellowish liquid, of spec. grav. 0.947, boiling at 16.6° (62° F.), and of an agreeable odor of apples. It is soluble in 48 parts of water, in all proportions of alcohol, and freely soluble in dilute alcohol. It gradually decomposes; more quickly in contact with water.

b. "Spirit of nitrous ether," of 5 per cent. nitrite of ethyl, boils at 63° C. (145° F.): the test-tube containing it being immersed in water of that temperature, and a few fragments of broken glass added.

Quantitative.—*c*. 10 grams of the spirit of nitrous ether are macerated, with 1.2 to 1.5 grams of fused potassa, in a stoppered flask, for 12 hours, occasionally agitating. Then pour the mixture into a beaker, dilute with an equal bulk of water, and leave at a warm temperature till the odor of alcohol disappears. Acidulate slightly with sulphuric acid, and add, from a burette, a solution of potassium permanganate of known strength, until the color ceases to be discharged. The number of grams of permanganate expended, multiplied by 1.18, equals the number of grams of ethyl nitrite in the 10 grams of material taken.

197. CHLOROFORM. $CHCl_3$. *Identified* by its sensible and physical properties (*a*); its liberation of chlorine when decomposed (*b*), and its production of isonitril (*c*). It acts as a reducing agent (*d*). It is *separated* by washing with concentrated sulphuric acid and with water, and rectification from alkaline carbonate, lime, calcium chloride, animal charcoal (*e*). It may be *estimated* from the chloride it gives after digestion with alcoholic fixed alkalies.

a. Chloroform is a colorless liquid, of spec. grav. 1.497 (at 15° C.), boiling at 61° C. (142° F.). It is neutral in reaction, and has an agreeable sweet ethereal odor and burning sweet taste. It is not readily combustible, but burns with paper, with a green-bordered flame. By standing, especially in the light and if free from alcohol, it becomes acid and gives reactions for chlorine and hydrochloric acid.—Chloroform is not miscible in water except in traces, but is *soluble* in all proportions of alcohol, ether, benzole, petroleum naphtha, bisulphide of carbon, fixed and volatile oils—not in concentrated sulphuric acid. It dissolves sulphur, phosphorus, iodine, iodoform, resins, caoutchouc, and gutta percha.

b. Chloroform is decomposed, with production of chlorine and hydrochloric acid, when it is passed in vapor through a red-hot tube; or, with production of chloride and formate, when digested with alcoholic solution of **potassa** (slowly by aqueous potassa). (Alcoholic ammonia produces ammonium cyanide and chloride—the better with help of potassa.)

$$CHCl_3 + 4KHO = 3KCl + KCHO_2 + 2H_2O$$
$$CHCl_3 + 5NH_3 = 3NH_4Cl + NH_4CN$$

Also, with production of hydrochloric acid, by nascent **hydrogen**, as evolved by zinc with sulphuric acid diluted with alcohol.—The free chlorine is made evident by potassic iodide (and starch), and the hydrochloric acid by silver salt. (Neither pure nor alcoholic chloroform affects silver nitrate.)

c. Chloroform, even in solution with 5,000 parts of alcohol, when treated with **anilin** (or other monamine) and then with **alcoholic soda**, forms an isonitril, recognized by its characteristic odor (HOFFMANN). This test distinguishes chloroform from Chlorœthylidene ($C_2H_4Cl_2$). Iodoform, Bromoform, Chloral, etc., react in the test, the same as chloroform.

d. Chloroform readily reduces the hot **potassio cupric** solution (distinction from chlorœthylidene and from alcohol).

e. Chloroform may be separated from slight mixtures of Ether, Alcohol, water, etc., as follows: To 10 parts of the impure chloroform, add 2 parts of concentrated sulphuric acid, and shake together occasionally for 24 hours. Remove the upper layer, add to it ½ part of (crystallized) carbonate of sodium previously dissolved in 1 part of water, agitate and digest (cold) for half an hour, then remove the lower layer and distil it from $\frac{1}{20}$ part of freshly-burned lime.—Distillation from dry calcium chloride separates chloroform from alcohol.—To separate from Ethereal Oil (ethyl and ethylene sulphates), distil from animal charcoal.

198. CHLORAL HYDRATE. $C_2HCl_3O.H_2O$. *Characterized* by its sensible and physical properties (*a*), and its formation of chloroform (*b*), and of chloralide (*c*). It has, with alkalies, considerable reducing power (*d*). *Separated* from chloral alcoholate by its slight solubility in cold chloroform and its greater solubility in cold water.—*Estimated* from the amount of chloroform it produces (*e*).

a. A friable solid, crystallizing from solvents in transparent rhomboidal crystals, or congealing in a white crystalline mass, melting at about 60° C. (140° F.) and boiling at 95° C. (203° F.)—(the Alcoholate boils at 116° C.). It slowly sublimes, in the bottle, at ordinary temperatures. It is neutral in reaction, and of an aromatic, penetrating, and slightly acrid odor, and bitter, caustic taste. Melted in a spoon, over the flame, it does not take fire (distinction from the alcoholate).—It is slightly deliquescent, readily soluble in 1½ parts of water (the alcoholate dissolves sparingly in cold water); soluble in alcohol, ether, benzole, petroleum naphtha, bisulphide of carbon; slightly soluble in cold chloroform (the alcoholate freely soluble). It forms liquid mixtures with camphor, and with phenic acid, and a crystalline mixture with glycerin.

b. Fixed and volatile **alkalies,** and their carbonates, in solution, decompose chloral hydrate—the chloroform subsiding from the milky mixture.

$$C_2HCl_3O.H_2O+KHO=CHCl_3+KCHO_2+H_2O$$

(100 parts chloral hydrate producing 72.2 parts of chloroform.) (Trichloracetic acid, also decomposed by alkalies into chloroform and formate, has an acid reaction, and boils at 195° C.)

c. Concentrated **sulphuric acid** separates, from about an equal weight of chloral hydrate, anhydrous chloral—the latter rising to the surface, as a pungent and irritating oily liquid, of spec. grav. 1.5.—Chloralide is formed when chloral hydrate (concentrated, if necessary, by distillation from chloride of calcium) is heated with about six times its volume of concentrated sulphuric acid, at 125° C., for some time. When cool, the mixture is diluted with six measures of water, and, if carbonized at all, extracted with ether. On evaporating the ether the chloralide ($C_5H_2Cl_6O_3$) crystallizes in stellate groups of prisms (or in needles) which melt at 116° C. and burn at 200° C. with a green-edged flame.—In certain conditions, sulphuric acid changes chloral into metachloral (insoluble in water, alcohol, or ether).

d. Chloral hydrate, in the act of decomposition by ammonia, promptly reduces **nitrate of silver** as a specular coating.—Aqueous solution of pure chloral hydrate does not within a few minutes perceptibly decolorize the **permanganate** of potassium solution, and does not at all affect argentic nitrate.—The potassio cupric solution is reduced by chloral according to 197, *d.*

Quantitative.—*e.* Take 10 grams of the chloral hydrate, dissolve in the least quantity of water, and add, in a graduated tube holding 20 c.c., ammonia enough to be a slight excess for the absolute chloral hydrate taken, according to the equation in *b* (5 c.c. of water of ammonia of spec. grav. 0.90). Stopper tightly in the tube, which should be nearly filled by the liquid, and leave until the subsident layer no longer increases—four to twelve hours. The c.c. of chloroform are multiplied by 1.5 for grams. Closer results are obtained by taking 50 grams chloral hydrate.

199. IODOFORM. CHI_3. A sulphur-yellow *solid*, crystallizing in hexagonal plates, stars, and rosettes; melting, at 115° to 120° C., with partial vaporization and partial decomposition into carbon, hydriodic acid, and iodine. It has a saffron-like odor, reminding of chloroform and of iodine, and a taste like the same substances, becoming unpleasantly strong of iodine.—It is *soluble* in 13,000 parts of water (to which it imparts a slight odor and taste), in 80 parts of cold or 12 parts of boiling alcohol of 80 per cent., in 20 parts of ether, and soluble in chloroform, bisulphide of carbon, fixed and volatile oils. The alcoholic solution is straw-yellow; the ether solution, gold-yellow; both solutions are neutral, and have a sweet-ethereal, burning taste and iodine-like after-taste.—It is difficultly and imperfectly decomposed by boiling aqueous potassa, but (WITTSTEIN) alcoholic potassa decomposes it, forming iodide and formate (see chloroform, *b*).

200. CROTON-CHLORAL HYDRATE. $C_4H_3Cl_3O$. The trichlorinated aldehyde of crotonic acid.—Thin, dazzling-white plates, melting at 78° C., volatile in steam at 100° C., boiling at 163°. It has a sweetish, melon-like taste, and its vapor irritates the eyes. It is sparingly soluble in cold, freely in hot water, and soluble in alcohol and in glycerin. **Potassa** decomposes it with formation of potassic chloride and formate and dichlorallylene ($C_3H_2Cl_2$).

201. AMYLIC ALCOHOL. $C_5H_{12}O$. *Characterized* by its sensible and physical properties (*a*); by its production of red sulphamylic acid (*b*); by its formation of odorous ethers (*c*).—It is *separated* from alcohol by fractional evaporation or distillation, or by adding water and extracting with ether (*d*); from water, in the slight proportions miscible, by adding petroleum naphtha or benzole, or by adding common salt (*e*).

a. A colorless and transparent *liquid*, of spec. grav. 0.816, boiling at 132–3° C. (270° F.), and having a sharp taste and a characteristic, pungent odor. Its vapor excites coughing, a few moments after it is inhaled.—It is soluble in about 40 parts of water, less soluble in solution of common salt, soluble in all proportions of alcohol, ether, chloroform, benzole, petroleum naphtha, fixed and volatile oils. It makes a slowly evanescent oil-spot upon paper.—It burns with a smoky flame.

b. When two parts of amylic alcohol are digested warm with three parts of concentrated **sulphuric acid**, sulphamylic acid, or amyl sulphuric acid is formed—having a red color and dissolving freely in water.

c. Distilled or digested hot with concentrated sulphuric acid and potassic acetate, the odor of "pear-oil" is developed —from formation of amyl acetate.—Distilled or digested with sulphuric acid and a little water and bichromate of potassium, the apple-odor of valeric aldehyde is first generated, and then the peculiar odor of valeric acid (42).

d. It is separated from (aqueous) ethylic alcohol, by adding an equal volume of pure ether, and then to the whole an equal volume (or enough) water to cause the ether to separate. The latter will contain most of the amylic alcohol. Benzole or petroleum naphtha may be used instead of ether.

e. If from 100 c.c. of commercial "fusel-oil" are slowly distilled 5 c.c., and this be agitated with a saturated solution of common salt, the separation of an oil-layer of 2.5 c.c. or over indicates that there is less than 15 per cent. of "proof spirit" in the fusel-oil taken.

202. "FUSEL-OIL" contains, besides amylic alcohol, small proportions of Butyric, Valeric, and volatile Fatty Acids, and of propylic and butyric alcohols.—In examination of spirits for fusel-oil, add 2 or 3 c.c. of potassa solution, to about 30 c.c. of the material, and evaporate by a gentle heat to dryness. Add 5 or 6 c.c. of sulphuric acid and nearly as much water:

when, if the acids in question are present, their *odor* will be apparent.*

203. NITRITE OF AMYL. $C_5H_{11}NO_2$. A light-yellowish liquid, darkening when heated, of spec. grav. 0.877, boiling at about 96° C. Its vapor has a reddish-yellow color. Its odor resembles that of ethyl nitrite.—Sulphuric acid (concentrated) decomposes it with explosive violence, sometimes with combustion. Alcoholic potassa decomposes it quickly, forming potassic nitrite; aqueous potassa decomposes it slowly.

* Farther, see Prescott's Examination of Alcoholic Liquors, New York, 1875.

INDEX.

SCIENTIFIC BOOKS

PUBLISHED BY

D. VAN NOSTRAND,

23 MURRAY STREET & 27 WARREN STREET,

NEW YORK.

Weisbach's Mechanics.

New and Revised Edition.

8vo. Cloth. $10.00.

A MANUAL OF THE MECHANICS OF ENGINEERING, and of the Construction of Machines. By JULIUS WEISBACH, PH. D. Translated from the fourth augmented and improved German edition, by ECKLEY B. COXE, A.M., Mining Engineer. Vol. I.—Theoretical Mechanics. 1,100 pages, and 902 wood-cut illustrations.

ABSTRACT OF CONTENTS.—Introduction to the Calculus—The General Principles of Mechanics—Phoronomics, or the Purely Mathematical Theory of Motion—Mechanics, or the General Physical Theory of Motion—Statics of Rigid Bodies—The Application of Statics to Elasticity and Strength—Dynamics of Rigid Bodies—Statics of Fluids—Dynamics of Fluids—The Theory of Oscillation, etc.

"The present edition is an entirely new work, greatly extended and very much improved. It forms a text-book which must find its way into the hands, not only of every student, but of every engineer who desires to refresh his memory or acquire clear ideas on doubtful points."—*Manufacturer and Builder.*

"We hope the day is not far distant when a thorough course of study and education as such shall be demanded of the practising engineer, and with this view we are glad to welcome this translation to our tongue and shores of one of the most able of the educators of Europe."—*The Technologist.*

Francis' Lowell Hydraulics.

Third Edition.

4to. Cloth. $15.00.

LOWELL HYDRAULIC EXPERIMENTS — being a Selection from Experiments on Hydraulic Motors, on the Flow of Water over Weirs, and in Open Canals of Uniform Rectangular Section, made at Lowell, Mass. By J. B. FRANCIS, Civil Engineer. Third edition, revised and enlarged, including many New Experiments on Gauging Water in Open Canals, and on the Flow through Submerged Orifices and Diverging Tubes. With 23 copperplates, beautifully engraved, and about 100 new pages of text.

The work is divided into parts. PART I., on hydraulic motors, includes ninety-two experiments on an improved Fourneyron Turbine Water-Wheel, of about two hundred horse-power, with rules and tables for the construction of similar motors; thirteen experiments on a model of a centre-vent water-wheel of the most simple design, and thirty-nine experiments on a centre-vent water-wheel of about two hundred and thirty horse-power.

PART II. includes seventy-four experiments made for the purpose of determining the form of the formula for computing the flow of water over weirs; nine experiments on the effect of back-water on the flow over weirs; eighty-eight experiments made for the purpose of determining the formula for computing the flow over weirs of regular or standard forms, with several tables of comparisons of the new formula with the results obtained by former experimenters; five experiments on the flow over a dam in which the crest was of the same form as that built by the Essex Company across the Merrimack River at Lawrence, Massachusetts; twenty-one experiments on the effect of observing the depths of water on a weir at different distances from the weir; an extensive series of experiments made for the purpose of determining rules for gauging streams of water in open canals, with tables for facilitating the same; and one hundred and one experiments on the discharge of water through submerged orifices and diverging tubes, the whole being fully illustrated by twenty-three double plates engraved on copper.

In 1855 the proprietors of the Locks and Canals on Merrimack River consented to the publication of the first edition of this work, which contained a selection of the most important hydraulic experiments made at Lowell up to that time. In this edition the principal hydraulic experiments made there, subsequent to 1855, have been added, including the important series above mentioned, for determining rules for the gauging the flow of water in open canals, and the interesting series on the flow through a submerged Venturi's tube, in which a larger flow was obtained than any we find recorded.

Francis on Cast-Iron Pillars.

8vo. Cloth. $2.00.

ON THE STRENGTH OF CAST-IRON PILLARS, with Tables for the use of Engineers, Architects, and Builders. By JAMES B. FRANCIS, Civil Engineer.

Merrill's Iron Truss Bridges.

Second Edition.

4to. Cloth. $5.00.

IRON TRUSS BRIDGES FOR RAILROADS. The Method of Calculating Strains in Trusses, with a careful comparison of the most prominent Trusses, in reference to economy in combination, etc., etc. By Brevet Colonel WILLIAM E. MERRILL, U.S.A., Major Corps of Engineers. Nine lithographed plates of illustrations.

"The work before us is an attempt to give a basis for sound reform in this feature of railroad engineering, by throwing 'additional light upon the method of calculating the maxima strains that can come upon any part of a bridge truss, and upon the manner of proportioning each part, so that it shall be as strong relatively to its own strains as any other part, and so that the entire bridge may be strong enough to sustain several times as great strains as the greatest that can come upon it in actual use.' "—*Scientific American.*

"The author has presented his views in a clear and intelligent manner, and the ingenuity displayed in coloring the figures so as to present certain facts to the eye forms no inappreciable part of the merits of the work. The reduction of the 'formulæ for obtaining the strength, volume, and weight of a cast-iron pillar under a strain of compression,' will be very acceptable to those who have occasion hereafter to make investigations involving these conditions. As a whole, the work has been well done."—*Railroad Gazette, Chicago.*

Humber's Strains in Girders.

18mo. Cloth. $2.50.

A HANDY BOOK FOR THE CALCULATION OF STRAINS IN GIRDERS and Similar Structures, and their Strength, consisting of Formulæ and Corresponding Diagrams, with numerous details for practical application. By WILLIAM HUMBER. Fully illustrated.

Shreve on Bridges and Roofs.

8vo, 87 wood-cut illustrations. Cloth. $5.00.

A TREATISE ON THE STRENGTH OF BRIDGES AND ROOFS—comprising the determination of Algebraic formulas for Strains in Horizontal, Inclined or Rafter, Triangular, Bowstring, Lenticular and other Trusses, from fixed and moving loads, with practical applications and examples, for the use of Students and Engineers. By SAMUEL H. SHREVE, A.M., Civil Engineer.

"On the whole, Mr. Shreve has produced a book which is the simplest, clearest, and at the same time, the most systematic and with the best mathematical reasoning of any work upon the same subject in the language."—*Railroad Gazette.*

"From the unusually clear language in which Mr. Shreve has given every statement, the student will have but himself to blame if he does not become thorough master of the subject."—*London Mining Journal.*

"Mr. Shreve has produced a work that must always take high rank as a text-book, * * * and no Bridge Engineer should be without it, as a valuable work of reference, and one that will frequently assist him out of difficulties."—*Franklin Institute Journal.*

The Kansas City Bridge.

4to. Cloth. $6.00

WITH AN ACCOUNT OF THE REGIMEN OF THE MISSOURI RIVER, and a description of the Methods used for Founding in that River. By O. CHANUTE, Chief Engineer, and GEORGE MORISON, Assistant Engineer. Illustrated with five lithographic views and twelve plates of plans.

Illustrations.

VIEWS.—View of the Kansas City Bridge, August 2, 1869. Lowering Caisson No. 1 into position. Caisson for Pier No. 4 brought into position. View of Foundation Works, Pier No. 4. Pier No. 1.

PLATES.—I. Map showing location of Bridge. II. Water Record—Cross Section of River—Profile of Crossing—Pontoon Protection. III. Water Deadener—Caisson No. 2—Foundation Works, Pier No. 3. IV. Foundation Works, Pier No. 4. V. Foundation Works, Pier No. 4. VI. Caisson No. 5—Sheet Piling at Pier No. 6—Details of Dredges—Pile Shoe—Beton Box. VII. Masonry—Draw Protection—False Works between Piers 3 and 4. VIII. Floating Derricks. IX. General Elevation—176 feet span. X. 248 feet span. XI. Plans of Draw. XII. Strain Diagrams.

Clarke's Quincy Bridge.

4to. Cloth. $7.50.

DESCRIPTION OF THE IRON RAILWAY Bridge across the Mississippi River at Quincy, Illinois. By THOMAS CURTIS CLARKE, Chief Engineer. Illustrated with twenty-one lithographed plans.

Illustrations.

PLATES.—General Plan of Mississippi River at Quincy, showing location of Bridge. II*a*. General Sections of Mississippi River at Quincy, showing location of Bridge. II*b*. General Sections of Mississippi River at Quincy, showing location of Bridge. III. General Sections of Mississippi River at Quincy, showing location of Bridge. IV. Plans of Masonry. V. Diagram of Spans, showing the Dimensions, Arrangement of Panels, etc. VI. Two hundred and fifty feet span, and details. VII. Three hundred and sixty feet Pivot Draw. VIII. Details of three hundred and sixty feet Draw. IX. Ice-Breakers, Foundations of Piers and Abutments, Water Table, and Curve of Deflections. X. Foundations of Pier 2, in Process of Construction. XI. Foundations of Pier 3, and its Protection. XII. Foundations of Pier 3, in Process of Construction, and Steam Dredge. XIII. Foundations of Piers 5 to 18, in Process of Construction. XIV. False Works, showing Process of Handling and Setting Stone. XV. False Works for Raising Iron Work of Superstructure. XVI. Steam Dredge used in Foundations 9 to 18. XVII. Single Bucket Dredge used in Foundations of Bay Piers. XVIII. Saws used for Cutting Piles under water. XIX. Sand Pump and Concrete Box. XX. Masonry Travelling Crane.

Whipple on Bridge Building.

8vo, Illustrated. Cloth. $4.00.

AN ELEMENTARY AND PRACTICAL TREATISE ON BRIDGE BUILDING. An enlarged and improved edition of the Author's original work. By S. WHIPPLE, C. E., Inventor of the Whipple Bridges, &c. Second Edition.

The design has been to develop from Fundamental Principles a system easy of comprehension, and such as to enable the attentive reader and student to judge understandingly for himself, as to the relative merits of different plans and combinations, and to adopt for use such as may be most suitable for the cases he may have to deal with.

It is hoped the work may prove an appropriate Text-Book upon the subject treated of, for the Engineering Student, and a useful manual for the Practicing Engineer and Bridge Builder.

Stoney on Strains.

New and Revised Edition, with numerous illustrations.

Royal 8vo, 664 pp. Cloth. $12.50.

THE THEORY OF STRAINS IN GIRDERS and Similar Structures, with Observations on the Application of Theory to Practice, and Tables of Strength and other Properties of Materials. By BINDON B. STONEY, B. A.

Roebling's Bridges.

Imperial folio. Cloth. $25.00.

LONG AND SHORT SPAN RAILWAY BRIDGES. By JOHN A. ROEBLING, C. E. Illustrated with large copperplate engravings of plans and views.

List of Plates

1. Parabolic Truss Railway Bridge. 2, 3, 4, 5, 6. Details of Parabolic Truss, with centre span 500 feet in the clear. 7. Plan and View of a Bridge over the Mississippi River, at St. Louis, for railway and common travel. 8, 9, 10, 11, 12. Details and View of St. Louis Bridge. 13. Railroad Bridge over the Ohio.

Diedrichs' Theory of Strains.

8vo. Cloth. $5.00.

A Compendium for the Calculation and Construction of Bridges, Roofs, and Cranes, with the Application of Trigonometrical Notes. Containing the most comprehensive information in regard to the Resulting Strains for a permanent Load, as also for a combined (Permanent and Rolling) Load. In two sections adapted to the requirements of the present time. By JOHN DIEDRICHS. Illustrated by numerous plates and diagrams.

"The want of a compact, universal and popular treatise on the Construction of Roofs and Bridges—especially one treating of the influence of a variable load—and the unsatisfactory essays of different authors on the subject, induced me to prepare this work."

Whilden's Strength of Materials.

12mo. Cloth. $2.00.

ON THE STRENGTH OF MATERIALS used in Engineering Construction. By J. K. Whilden.

Campin on Iron Roofs.

Large 8vo. Cloth. $2.00.

ON THE CONSTRUCTION OF IRON ROOFS. A Theoretical and Practical Treatise. By Francis Campin. With wood-cuts and plates of Roofs lately executed.

"The mathematical formulas are of an elementary kind, and the process admits of an easy extension so as to embrace the prominent varieties of iron truss bridges. The treatise, though of a practical scientific character, may be easily mastered by any one familiar with elementary mechanics and plane trigonometry."

Holley's Railway Practice.

1 vol. folio. Cloth. $12.00.

AMERICAN AND EUROPEAN RAILWAY PRACTICE, in the Economical Generation of Steam, including the materials and construction of Coal-burning Boilers, Combustion, the Variable Blast, Vaporization, Circulation, Super-heating, Supplying and Heating Feed-water, &c., and the adaptation of Wood and Coke-burning Engines to Coal-burning; and in Permanent Way, including Road-bed, Sleepers, Rails, Joint Fastenings, Street Railways, &c., &c. By Alexander L. Holley, B. P. With 77 lithographed plates.

"This is an elaborate treatise by one of our ablest civil engineers, on the construction and use of locomotives, with a few chapters on the building of Railroads. * * * All these subjects are treated by the author, who is a first-class railroad engineer, in both an intelligent and intelligible manner. The facts and ideas are well arranged, and presented in a clear and simple style, accompanied by beautiful engravings, and we presume the work will be regarded as indispensable by all who are interested in a knowledge of the construction of railroads and rolling stock, or the working of locomotives."—*Scientific American.*

Henrici's Skeleton Structures.

8vo. Cloth. $1.50.

SKELETON STRUCTURES, especially in their Application to the building of Steel and Iron Bridges. By Olaus Henrici. With folding plates and diagrams.

By presenting these general examinations on Skeleton Structures, with particular application for Suspended Bridges, to Engineers, I venture to express the hope that they will receive these theoretical results with some confidence, even although an opportunity is wanting to compare them with practical results. O. H.

Useful Information for Railway Men.

Pocket form. Morocco, gilt, $2.00.

Compiled by W. G. Hamilton, Engineer. Fifth edition, revised and enlarged. 570 pages.

"It embodies many valuable formulæ and recipes useful for railway men, and, indeed, for almost every class of persons in the world. The 'information' comprises some valuable formulæ and rules for the construction of boilers and engines, masonry, properties of steel and iron, and the strength of materials generally."—*Railroad Gazette, Chicago.*

Brooklyn Water Works.

1 vol. folio. Cloth. $25.00.

A DESCRIPTIVE ACCOUNT OF THE CONSTRUCTION OF THE WORKS, and also Reports on the Brooklyn, Hartford, Belleville, and Cambridge Pumping Engines. Prepared and printed by order of the Board of Water Commissioners. With 59 illustrations.

Contents.—Supply Ponds—The Conduit—Ridgewood Engine House and Pump Well—Ridgewood Engines—Force Mains—Ridgewood Reservoir—Pipe Distribution—Mount Prospect Reservoir—Mount Prospect Engine House and Engine—Drainage Grounds—Sewerage Works—Appendix.

Kirkwood on Filtration.

4to. Cloth. $15.00.

REPORT ON THE FILTRATION OF RIVER WATERS, for the Supply of Cities, as practised in Europe, made to the Board of Water Commissioners of the City of St. Louis. By JAMES P. KIRKWOOD. Illustrated by 30 double-plate engravings.

CONTENTS.—Report on Filtration—London Works, General—Chelsea Water Works and Filters—Lambeth Water Works and Filters—Southwark and Vauxhall Water Works and Filters—Grand Junction Water Works and Filters—West Middlesex Water Works and Filters—New River Water Works and Filters—East London Water Works and Filters—Leicester Water Works and Filters—York Water Works and Filters—Liverpool Water Works and Filters—Edinburgh Water Works and Filters—Dublin Water Works and Filters—Perth Water Works and Filtering Gallery—Berlin Water Works and Filters—Hamburg Water Works and Reservoirs—Altona Water Works and Filters—Tours Water Works and Filtering Canal—Angers Water Works and Filtering Galleries—Nantes Water Works and Filters—Lyons Water Works and Filtering Galleries—Toulouse Water Works and Filtering Galleries—Marseilles Water Works and Filters—Genoa Water Works and Filtering Galleries—Leghorn Water Works and Cisterns—Wakefield Water Works and Filters—Appendix.

Tunner on Roll-Turning.

1 vol. 8vo. and 1 vol. plates. $10.00.

A TREATISE ON ROLL-TURNING FOR THE MANUFACTURE OF IRON. By PETER TUNNER. Translated and adapted. By JOHN B. PEARSE, of the Pennsylvania Steel Works. With numerous wood-cuts, 8vo., together with a folio atlas of 10 lithographed plates of Rolls, Measurements, &c.

"We commend this book as a clear, elaborate, and practical treatise upon the department of iron manufacturing operations to which it is devoted. The writer states in his preface, that for twenty-five years he has felt the necessity of such a work, and has evidently brought to its preparation the fruits of experience, a painstaking regard for accuracy of statement, and a desire to furnish information in a style readily understood. The book should be in the hands of every one interested, either in the general practice of mechanical engineering, or the special branch of manufacturing operations to which the work relates."—*American Artisan.*

Glynn on the Power of Water.

12mo. Cloth. $1.00.

A TREATISE ON THE POWER OF WATER, as applied to drive Flour Mills, and to give motion to Turbines and other Hydrostatic Engines. By JOSEPH GLYNN, F.R. S. Third edition, revised and enlarged, with numerous illustrations.

Hewson on Embankments.

8vo. Cloth. $2.00.

PRINCIPLES AND PRACTICE OF EMBANKING LANDS from River Floods, as applied to the Levees of the Mississippi. By WILLIAM HEWSON, Civil Engineer.

"This is a valuable treatise on the principles and practice of embanking lands from river floods, as applied to the Levees of the Mississippi, by a highly intelligent and experienced engineer. The author says it is a first attempt to reduce to order and to rule the design, execution, and measurement of the Levees of the Mississippi. It is a most useful and needed contribution to scientific literature.—*Philadelphia Evening Journal.*

Grüner on Steel.

8vo. Cloth. $3.50.

THE MANUFACTURE OF STEEL. By M. L. GRUNER, translated from the French. By Lenox Smith, A. M., E. M., with an appendix on the Bessemer Process in the United States, by the translator. Illustrated by lithographed drawings and wood-cuts.

"The purpose of the work is to present a careful, elaborate, and at the same time practical examination into the physical properties of steel, as well as a description of the new processes and mechanical appliances for its manufacture. The information which it contains, gathered from many trustworthy sources, will be found of much value to the American steel manufacturer, who may thus acquaint himself with the results of careful and elaborate experiments in other countries, and better prepare himself for successful competition in this important industry with foreign makers. The fact that this volume is from the pen of one of the ablest metallurgists of the present day, cannot fail, we think, to secure for it a favorable consideration.—*Iron Age.*

Bauerman on Iron.

12mo. Cloth. $2.00.

TREATISE ON THE METALLURGY OF IRON. Containing outlines of the History of Iron Manufacture, methods of Assay, and analysis of Iron Ores, processes of manufacture of Iron and Steel, etc., etc. By H. Bauerman. First American edition. Revised and enlarged, with an appendix on the Martin Process for making Steel, from the report of Abram S. Hewitt. Illustrated with numerous wood engravings.

"This is an important addition to the stock of technical works published in this country. It embodies the latest facts, discoveries, and processes connected with the manufacture of iron and steel, and should be in the hands of every person interested in the subject, as well as in all technical and scientific libraries."—*Scientific American.*

Link and Valve Motions, by W. S. Auchincloss.

8vo. Cloth. $3.00.

APPLICATION OF THE SLIDE VALVE and Link Motion to Stationary, Portable, Locomotive and Marine Engines, with new and simple methods for proportioning the parts. By William S. Auchincloss, Civil and Mechanical Engineer. Designed as a hand-book for Mechanical Engineers, Master Mechanics, Draughtsmen and Students of Steam Engineering. All dimensions of the valve are found with the greatest ease by means of a Printed Scale, and proportions of the link determined *without* the assistance of a model. Illustrated by 37 wood-cuts and 21 lithographic plates, together with a copperplate engraving of the Travel Scale.

All the matters we have mentioned are treated with a clearness and absence of unnecessary verbiage which renders the work a peculiarly valuable one. The Travel Scale only requires to be known to be appreciated. Mr. A. writes so ably on his subject, we wish he had written more. *London Engineering.*

We have never opened a work relating to steam which seemed to us better calculated to give an intelligent mind a clear understanding of the department it discusses.—*Scientific American.*

Slide Valve by Eccentrics, by Prof. C. W. MacCord.

4to. Illustrated. Cloth, $4.00.

A PRACTICAL TREATISE ON THE SLIDE VALVE BY ECCENTRICS, examining by methods, the action of the Eccentric upon the Slide Valve, and explaining the practical processes of laying out the movements, adapting the valve for its various duties in the steam-engine. For the use of Engineers, Draughtsmen, Machinists, and Students of valve motions in general. By C. W. MacCord, A. M., Professor of Mechanical Drawing, Stevens' Institute of Technology, Hoboken, N. J.

Stillman's Steam-Engine Indicator.

12mo. Cloth. $1.00.

THE STEAM-ENGINE INDICATOR, and the Improved Manometer Steam and Vacuum Gauges; their utility and application By Paul Stillman. New edition.

Bacon's Steam-Engine Indicator.

12mo. Cloth. $1.00. Mor. $1.50.

A TREATISE ON THE RICHARDS STEAM-ENGINE INDICATOR, with directions for its use. By Charles T. Porter. Revised, with notes and large additions as developed by American Practice, with an Appendix containing useful formulæ and rules for Engineers. By F. W. Bacon, M. E., Member of the American Society of Civil Engineers. Illustrated. Second Edition

In this work, Mr. Porter's book has been taken as the basis, but Mr. Bacon has adapted it to American Practice, and has conferred a great boon on American Engineers.—*Artisan.*

Bartol on Marine Boilers.

8vo. Cloth. $1.50.

TREATISE ON THE MARINE BOILERS OF THE UNITED STATES. By H. B. Bartol. Illustrated.

Gillmore's Limes and Cements.

Fourth Edition. Revised and Enlargd.

8vo. Cloth. $4.00.

PRACTICAL TREATISE ON LIMES, HYDRAULIC CEMENTS, AND MORTARS. Papers on Practical Engineering, U. S. Engineer Department, No. 9, containing Reports of numerous experiments conducted in New York City, during the years 1858 to 1861, inclusive. By Q. A. GILLMORE, Brig-General U. S. Volunteers, and Major U. S. Corps of Engineers. With numerous illustrations.

"This work contains a record of certain experiments and researches made under the authority of the Engineer Bureau of the War Department from 1858 to 1861, upon the various hydraulic cements of the United States, and the materials for their manufacture. The experiments were carefully made, and are well reported and compiled."—*Journal Franklin Institute.*

Gillmore's Coignet Beton.

8vo. Cloth. $2.50.

COIGNET BETON AND OTHER ARTIFICIAL STONE. By Q. A. GILLMORE. 9 Plates, Views, etc.

This work describes with considerable minuteness of detail the several kinds of artificial stone in most general use in Europe and now beginning to be introduced in the United States, discusses their properties, relative merits, and cost, and describes the materials of which they are composed. . . . . The subject is one of special and growing interest, and we commend the work, embodying as it does the matured opinions of an experienced engineer and expert.

Williamson's Practical Tables.

4to. Flexible Cloth. $2.50.

PRACTICAL TABLES IN METEOROLOGY AND HYPSOMETRY, in connection with the use of the Barometer. By Col. R. S. WILLIAMSOM, U. S. A.

Williamson on the Barometer.

4to. Cloth. $15.00.

ON THE USE OF THE BAROMETER ON SURVEYS AND RECONNAISSANCES. Part I. Meteorology in its Connection with Hypsometry. Part II. Barometric Hypsometry. By R. S. WILLIAMSON, Bvt. Lieut.-Col. U. S. A., Major Corps of Engineers. With Illustrative Tables and Engravings. Paper No. 15, Professional Papers, Corps of Engineers.

"SAN FRANCISCO, CAL., *Feb.* 27, 1867.

"Gen. A. A. HUMPHREYS, Chief of Engineers, U. S. Army:

"GENERAL,—I have the honor to submit to you, in the following pages, the results of my investigations in meteorology and hypsometry, made with the view of ascertaining how far the barometer can be used as a reliable instrument for determining altitudes on extended lines of survey and reconnaissances. These investigations have occupied the leisure permitted me from my professional duties during the last ten years, and I hope the results will be deemed of sufficient value to have a place assigned them among the printed professional papers of the United States Corps of Engineers.

"Very respectfully, your obedient servant,

"R. S. WILLIAMSON,

"Bvt. Lt.-Col. U. S. A., Major Corps of U. S. Engineers."

Von Cotta's Ore Deposits.

8vo. Cloth. $4.00.

TREATISE ON ORE DEPOSITS. By BERNHARD VON COTTA, Professor of Geology in the Royal School of Mines, Freidberg, Saxony. Translated from the second German edition, by FREDERICK PRIME, Jr., Mining Engineer, and revised by the author, with numerous illustrations.

"Prof. Von Cotta of the Freiberg School of Mines, is the author of the best modern treatise on ore deposits, and we are heartily glad that this admirable work has been translated and published in this country. The translator, Mr. Frederick Prime, Jr., a graduate of Freiberg, has had in his work the great advantage of a revision by the author himself, who declares in a prefatory note that this may be considered as a new edition (the third) of his own book.

"It is a timely and welcome contribution to the literature of mining in this country, and we are grateful to the translator for his enterprise and good judgment in undertaking its preparation; while we recognize with equal cordiality the liberality of the author in granting both permission and assistance."—*Extract from Review in Engineering and Mining Journal.*

Plattner's Blow-Pipe Analysis.

Second edition. Revised. 8vo. Cloth. $7.50.

PLATTNER'S MANUAL OF QUALITATIVE AND QUANTITATIVE ANALYSIS WITH THE BLOW-PIPE. From the last German edition Revised and enlarged. By Prof. TH. RICHTER, of the Royal Saxon Mining Academy. Translated by Prof. H. B. CORNWALL, Assistant in the Columbia School of Mines, New York; assisted by JOHN H. CASWELL. Illustrated with eighty-seven wood-cuts and one Lithographic Plate. 560 pages.

"Plattner's celebrated work has long been recognized as the only complete book on Blow-Pipe Analysis. The fourth German edition, edited by Prof. Richter, fully sustains the reputation which the earlier editions acquired during the lifetime of the author, and it is a source of great satisfaction to us to know that Prof. Richter has co-operated with the translator in issuing the American edition of the work, which is in fact a fifth edition of the original work, being far more complete than the last German edition."—*Silliman's Journal.*

There is nothing so complete to be found in the English language. Plattner's book is not a mere pocket edition; it is intended as a comprehensive guide to all that is at present known on the blow-pipe, and as such is really indispensable to teachers and advanced pupils.

"Mr. Cornwall's edition is something more than a translation, as it contains many corrections, emendations and additions not to be found in the original. It is a decided improvement on the work in its German dress."—*Journal of Applied Chemistry.*

Egleston's Mineralogy.

8vo. Illustrated with 34 Lithographic Plates. Cloth. $4.50.

LECTURES ON DESCRIPTIVE MINERALOGY, Delivered at the School of Mines, Columbia College. BY PROFESSOR T. EGLESTON.

These lectures are what their title indicates, the lectures on Mineralogy delivered at the School of Mines of Columbia College. They have been printed for the students, in order that more time might be given to the various methods of examining and determining minerals. The second part has only been printed. The first part, comprising crystallography and physical mineralogy, will be printed at some future time.

Pynchon's Chemical Physics.

New Edition. Revised and Enlarged.

Crown 8vo. Cloth. $3.00.

INTRODUCTION TO CHEMICAL PHYSICS, Designed for the Use of Academies, Colleges, and High Schools. Illustrated with numerous engravings, and containing copious experiments with directions for preparing them. By THOMAS RUGGLES PYNCHON, M.A., Professor of Chemistry and the Natural Sciences, Trinity College, Hartford.

Hitherto, no work suitable for general use, treating of all these subjects within the limits of a single volume, could be found; consequently the attention they have received has not been at all proportionate to their importance. It is believed that a book containing so much valuable information within so small a compass, cannot fail to meet with a ready sale among all intelligent persons, while Professional men, Physicians, Medical Students, Photographers, Telegraphers, Engineers, and Artisans generally, will find it specially valuable, if not nearly indispensable, as a book of reference.

"We strongly recommend this able treatise to our readers as the first work ever published on the subject free from perplexing technicalities. In style it is pure, in description graphic, and its typographical appearance is artistic. It is altogether a most excellent work."—*Eclectic Medical Journal.*

"It treats fully of Photography, Telegraphy, Steam Engines, and the various applications of Electricity. In short, it is a carefully prepared volume, abreast with the latest scientific discoveries and inventions."—*Hartford Courant.*

Plympton's Blow-Pipe Analysis.

12mo. Cloth. $1 50.

THE BLOW-PIPE: A Guide to Its Use in the Determination of Salts and Minerals. Compiled from various sources, by GEORGE W. PLYMPTON, C.E., A.M., Professor of Physical Science in the Polytechnic Institute, Brooklyn, N. Y.

"This manual probably has no superior in the English language as a textbook for beginners, or as a guide to the student working without a teacher. To the latter many illustrations of the utensils and apparatus required in using the blow-pipe, as well as the fully illustrated description of the blow-pipe flame, will be especially serviceable."—*New York Teacher.*

Ure's Dictionary.

Sixth Edition.

London, 1872.

3 vols. 8vo. Cloth, $25.00. Half Russia, $32.50.

DICTIONARY OF ARTS, MANUFACTURES, AND MINES. By ANDREW URE, M.D. Sixth edition. Edited by ROBERT HUNT, F.R.S., greatly enlarged and rewritten.

Brande and Cox's Dictionary.

New Edition.

London, 1872.

3 vols. 8vo. Cloth, $20.00. Half Morocco, $27.50.

A Dictionary of Science, Literature, and Art. Edited by W. T. BRANDE and Rev. GEO. W. COX. New and enlarged edition.

Watt's Dictionary of Chemistry.

Supplementary Volume.

8vo. Cloth. $9.00.

This volume brings the Record of Chemical Discovery down to the end of the year 1869, including also several additions to, and corrections of, former results which have appeared in 1870 and 1871.

*** Complete Sets of the Work, New and Revised edition, including above supplement. 6 vols. 8vo. Cloth. $62.00.

Rammelsberg's Chemical Analysis.

8vo. Cloth. $2.25.

GUIDE TO A COURSE OF QUANTITATIVE CHEMICAL ANALYSIS, ESPECIALLY OF MINERALS AND FURNACE PRODUCTS. Illustrated by Examples. By C. F. RAMMELSBERG. Translated by J. TOWLER, M.D.

This work has been translated, and is now published expressly for those students in chemistry whose time and other studies in colleges do not permit them to enter upon the more elaborate and expensive treatises of Fresenius and others. It is the condensed labor of a master in chemistry and of a practical analyst.

Eliot and Storer's Qualitative Chemical Analysis.

New Edition, Revised.

12mo. Illustrated. Cloth. $1.50.

A COMPENDIOUS MANUAL OF QUALITATIVE CHEMICAL ANALYSIS. By CHARLES W. ELIOT and FRANK H. STORER. Revised with the Coöperation of the Authors, by WILLIAM RIPLEY NICHOLS, Professor of Chemistry in the Massachusetts Institute of Technology.

"This Manual has great merits as a practical introduction to the science and the art of which it treats. It contains enough of the theory and practice of qualitative analysis, "in the wet way," to bring out all the reasoning involved in the science, and to present clearly to the student the most approved methods of the art. It is specially adapted for exercises and experiments in the laboratory; and yet its classifications and manner of treatment are so systematic and logical throughout, as to adapt it in a high degree to that higher class of students generally who desire an accurate knowledge of the practical methods of arriving at scientific facts."—*Lutheran Observer.*

"We wish every academical class in the land could have the benefit of the fifty exercises of two hours each necessary to master this book. Chemistry would cease to be a mere matter of memory, and become a pleasant experimental and intellectual recreation. We heartily commend this little volume to the notice of those teachers who believe in using the sciences as means of mental discipline."—*College Courant.*

Craig's Decimal System.

Square 32mo. Limp. 50c.

WEIGHTS AND MEASURES. An Account of the Decimal System, with Tables of Conversion for Commercial and Scientific Uses. By B. F. CRAIG, M. D.

"The most lucid, accurate, and useful of all the hand-books on this subject that we have yet seen. It gives forty-seven tables of comparison between the English and French denominations of length, area, capacity, weight, and the Centigrade and Fahrenheit thermometers, with clear instructions how to use them; and to this practical portion, which helps to make the transition as easy as possible, is prefixed a scientific explanation of the errors in the metric system, and how they may be corrected in the laboratory."—*Nation.*

Nugent on Optics.

12mo. Cloth. $2.00

TREATISE ON OPTICS; or, Light and Sight, theoretically and practically treated; with the application to Fine Art and Industrial Pursuits. By E. Nugent. With one hundred and three illustrations.

"This book is of a practical rather than a theoretical kind, and is designed to afford accurate and complete information to all interested in applications of the science."—*Round Table.*

Barnard's Metric System.

8vo. Brown cloth. $3.00.

THE METRIC SYSTEM OF WEIGHTS AND MEASURES. An Address delivered before the Convocation of the University of the State of New York, at Albany, August, 1871. By Frederick A. P. Barnard, President of Columbia College, New York City. Second edition from the Revised edition printed for the Trustees of Columbia College. Tinted paper.

"It is the best summary of the arguments in favor of the metric weights and measures with which we are acquainted, not only because it contains in small space the leading facts of the case, but because it puts the advocacy of that system on the only tenable grounds, namely, the great convenience of a decimal notation of weight and measure as well as money, the value of international uniformity in the matter, and the fact that this metric system is adopted and in general use by the majority of civilized nations."—*The Nation.*

The Young Mechanic.

Illustrated. 12mo. Cloth. $1.75.

THE YOUNG MECHANIC. Containing directions for the use of all kinds of tools, and for the construction of steam engines and mechanical models, including the Art of Turning in Wood and Metal. By the author of "The Lathe and its Uses," etc. From the English edition, with corrections.

Harrison's Mechanic's Tool-Book.

12mo. Cloth. $1.50.

MECHANIC'S TOOL BOOK, with practical rules and suggestions, for the use of Machinists, Iron Workers, and others. By W. B. HARRISON, Associate Editor of the "American Artisan." Illustrated with 44 engravings.

"This work is specially adapted to meet the wants of Machinists and workers in iron generally. It is made up of the work-day experience of an intelligent and ingenious mechanic, who had the faculty of adapting tools to various purposes. The practicability of his plans and suggestions are made apparent even to the unpractised eye by a series of well-executed wood engravings."—*Philadelphia Inquirer.*

Pope's Modern Practice of the Electric Telegraph.

Eighth Edition. 8vo. Cloth $2.00.

A Hand-book for Electricians and Operators. By FRANK L. POPE. Seventh edition. Revised and enlarged, and fully illustrated.

Extract from Letter of Prof. Morse.

"I have had time only cursorily to examine its contents, but this examination has resulted in great gratification, especially at the fairness and unprejudiced tone of your whole work.

"Your illustrated diagrams are admirable and beautifully executed.

"I think all your instructions in the use of the telegraph apparatus judicious and correct, and I most cordially wish you success."

Extract from Letter of Prof. G. W. Hough, of the Dudley Observatory.

"There is no other work of this kind in the English language that contains in so small a compass so much practical information in the application of galvanic electricity to telegraphy. It should be in the hands of every one interested in telegraphy, or the use of Batteries for other purposes."

Morse's Telegraphic Apparatus.

Illustrated. 8vo. Cloth. $2.00.

EXAMINATION OF THE TELEGRAPHIC APPARATUS AND THE PROCESSES IN TELEGAPHY. By SAMUEL F. B. MORSE, LL.D., United States Commissioner Paris Universal Exposition, 1867.

Sabine's History of the Telegraph.

12mo. Cloth. $1.25.

HISTORY AND PROGRESS OF THE ELECTRIC TELEGRAPH, with Descriptions of some of the Apparatus. By ROBERT SABINE, C. E. Second edition, with additions.

CONTENTS.—I. Early Observations of Electrical Phenomena. II. Telegraphs by Frictional Electricity. III. Telegraphs by Voltaic Electricity. IV. Telegraphs by Electro-Magnetism and Magneto-Electricity. V. Telegraphs now in use. VI. Overhead Lines. VII. Submarine Telegraph Lines. VIII. Underground Telegraphs. IX. Atmospheric Electricity.

Haskins' Galvanometer.

Pocket form. Illustrated. Morocco tucks. $2.00.

THE GALVANOMETER, AND ITS USES; a Manual for Electricians and Students. By C. H. HASKINS.

"We hope this excellent little work will meet with the sale its merits entitle it to. To every telegrapher who owns, or uses a Galvanometer, or ever expects to, it will be quite indispensable."—*The Telegrapher.*

Culley's Hand-Book of Telegraphy.

8vo. Cloth. $5.00.

A HAND-BOOK OF PRACTICAL TELEGRAPHY. By R. S. CULLEY, Engineer to the Electric and International Telegraph Company. Fifth edition, revised and enlarged.

Foster's Submarine Blasting.

4to. Cloth. $3.50.

SUBMARINE BLASTING in Boston Harbor, Massachusetts—Removal of Tower and Corwin Rocks. By JOHN G. FOSTER, Lieutenant-Colonel of Engineers, and Brevet Major-General, U. S. Army. Illustrated with seven plates.

LIST OF PLATES.—1. Sketch of the Narrows, Boston Harbor. 2. Townsend's Submarine Drilling Machine, and Working Vessel attending. 3. Submarine Drilling Machine employed. 4. Details of Drilling Machine employed. 5. Cartridges and Tamping used. 6. Fuses and Insulated Wires used. 7. Portable Friction Battery used.

Barnes' Submarine Warfare.

8vo. Cloth. $5.00.

SUBMARINE WARFARE, DEFENSIVE AND OFFENSIVE. Comprising a full and complete History of the Invention of the Torpedo, its employment in War and results of its use. Descriptions of the various forms of Torpedoes, Submarine Batteries and Torpedo Boats actually used in War. Methods of Ignition by Machinery, Contact Fuzes, and Electricity, and a full account of experiments made to determine the Explosive Force of Gunpowder under Water. Also a discussion of the Offensive Torpedo system, its effect upon Iron-Clad Ship systems, and influence upon Future Naval Wars. By Lieut.-Commander JOHN S. BARNES, U. S. N. With twenty lithographic plates and many wood-cuts.

"A book important to military men, and especially so to engineers and artillerists. It consists of an examination of the various offensive and defensive engines that have been contrived for submarine hostilities, including a discussion of the torpedo system, its effects upon iron-clad ship-systems, and its probable influence upon future naval wars. Plates of a valuable character accompany the treatise, which affords a useful history of the momentous subject it discusses. A great deal of useful information is collected in its pages, especially concerning the inventions of SCHOLL and VERDU, and of JONES' and HUNT'S batteries, as well as of other similar machines, and the use in submarine operations of gun-cotton and nitro-glycerine."—*N. Y. Times.*

Randall's Quartz Operator's Hand-Book.

12mo. Cloth. $2.00.

QUARTZ OPERATOR'S HAND-BOOK. By P. M. RANDALL. New edition, revised and enlarged. Fully illustrated.

The object of this work has been to present a clear and comprehensive exposition of mineral veins, and the means and modes chiefly employed for the mining and working of their ores—more especially those containing gold and silver.

Mitchell's Manual of Assaying.

8vo. Cloth. $10.00.

A MANUAL OF PRACTICAL ASSAYING. By JOHN MITCHELL. Third edition. Edited by WILLIAM CROOKES, F.R.S.

In this edition are incorporated all the late important discoveries in Assaying made in this country and abroad, and special care is devoted to the very important Volumetric and Colorimetric Assays, as well as to the Blow-Pipe Assays.

Benét's Chronoscope.

Second Edition.

Illustrated. 4to. Cloth. $3.00.

ELECTRO-BALLISTIC MACHINES, and the Schultz Chronoscope. By Lieutenant-Colonel S. V. BENÉT, Captain of Ordnance, U. S. Army.

CONTENTS.—1. Ballistic Pendulum. 2. Gun Pendulum. 3. Use of Electricity. 4. Navez' Machine. 5. Vignotti's Machine, with Plates. 6. Benton's Electro-Ballistic Pendulum, with Plates. 7. Leur's Tro-Pendulum Machine 8. Schultz's Chronoscope, with two Plates.

Michaelis' Chronograph.

4to. Illustrated. Cloth. $3.00.

THE LE BOULENGÉ CHRONOGRAPH. With three lithographed folding plates of illustrations. By Brevet Captain O E. MICHAELIS, First Lieutenant Ordnance Corps, U. S. Army.

"The excellent monograph of Captain Michaelis enters minutely into the details of construction and management, and gives tables of the times of flight calculated upon a given fall of the chronometer for all distances. Captain Michaelis has done good service in presenting this work to his brother officers, describing, as it does, an instrument which bids fair to be in constant use in our future ballistic experiments."—*Army and Navy Journal.*

Silversmith's Hand-Book.

Fourth Edition.

Illustrated. 12mo. Cloth. $3.00.

A PRACTICAL HAND-BOOK FOR MINERS, Metallurgists, and Assayers, comprising the most recent improvements in the disintegration, amalgamation, smelting, and parting of the Precious Ores, with a Comprehensive Digest of the Mining Laws. Greatly augmented, revised, and corrected. By JULIUS SILVERSMITH. Fourth edition. Profusely illustrated. 1 vol. 12mo. Cloth. $3.00.

One of the most important features of this work is that in which the metallurgy of the precious metals is treated of. In it the author has endeavored to embody all the processes for the reduction and manipulation of the precious ores heretofore successfully employed in Germany, England, Mexico, and the United States, together with such as have been more recently invented, and not yet fully tested—all of which are profusely illustrated and easy of comprehension.

Simms' Levelling.

8vo. Cloth. $2.50.

A TREATISE ON THE PRINCIPLES AND PRACTICE OF LEVELLING, showing its application to purposes of Railway Engineering and the Construction of Roads, &c. By FREDERICK W. SIMMS, C. E. From the fifth London edition, revised and corrected, with the addition of Mr. Law's Practical Examples for Setting Out Railway Curves. Illustrated with three lithographic plates and numerous wood-cuts.

" One of the most important text-books for the general surveyor, and there is scarcely a question connected with levelling for which a solution would be sought, but that would be satisfactorily answered by consulting this volume." —*Mining Journal.*

"The text-book on levelling in most of our engineering schools and colleges."—*Engineers.*

"The publishers have rendered a substantial service to the profession, especially to the younger members, by bringing out the present edition of Mr. Simms' useful work."—*Engineering.*

Stuart's Successful Engineer.

18mo. Boards. 50 cents.

HOW TO BECOME A SUCCESSFUL ENGINEER: Being Hints to Youths intending to adopt the Profession. By BERNARD STUART, Engineer. Sixth Edition.

"A valuable little book of sound, sensible advice to young men who wish to rise in the most important of the professions."—*Scientific American.*

Stuart's Naval Dry Docks.

Twenty-four engravings on steel.

Fourth Edition.

4to. Cloth. $6.00.

THE NAVAL DRY DOCKS OF THE UNITED STATES. By CHARLES B. STUART. Engineer in Chief of the United States Navy.

List of Illustrations.

Pumping Engine and Pumps—Plan of Dry Dock and Pump-Well—Sections of Dry Dock—Engine House—Iron Floating Gate—Details of Floating Gate—Iron Turning Gate—Plan of Turning Gate—Culvert Gate—Filling Culvert Gates—Engine Bed—Plate, Pumps, and Culvert—Engine House Roof—Floating Sectional Dock—Details of Section, and Plan of Turn-Tables—Plan of Basin and Marine Railways—Plan of Sliding Frame, and Elevation of Pumps—Hydraulic Cylinder—Plan of Gearing for Pumps and End Floats—Perspective View of Dock, Basin, and Railway—Plan of Basin of Portsmouth Dry Dock—Floating Balance Dock—Elevation of Trusses and the Machinery—Perspective View of Balance Dry Dock

---

Free Hand Drawing.

Profusely Illustrated. 18mo. Boards. 50 cents.

A GUIDE TO ORNAMENTAL, Figure, and Landscape Drawing. By an Art Student.

CONTENTS.—Materials employed in Drawing, and how to use them—On Lines and how to Draw them—On Shading—Concerning lines and shading, with applications of them to simple elementary subjects—Sketches from Nature.

Minifie's Mechanical Drawing.

Eighth Edition.

Royal 8vo. Cloth. $4.00.

A TEXT-BOOK OF GEOMETRICAL DRAWING for the use of Mechanics and Schools, in which the Definitions and Rules of Geometry are familiarly explained; the Practical Problems are arranged, from the most simple to the more complex, and in their description technicalities are avoided as much as possible. With illustrations for Drawing Plans, Sections, and Elevations of Buildings and Machinery; an Introduction to Isometrical Drawing, and an Essay on Linear Perspective and Shadows. Illustrated with over 200 diagrams engraved on steel. By WM. MINIFIE, Architect. Eighth Edition. With an Appendix on the Theory and Application of Colors.

"It is the best work on Drawing that we have ever seen, and is especially a text-book of Geometrical Drawing for the use of Mechanics and Schools. No young Mechanic, such as a Machinist, Engineer, Cabinet-Maker, Millwright, or Carpenter, should be without it."—*Scientific American.*

"One of the most comprehensive works of the kind ever published, and cannot but possess great value to builders. The style is at once elegant and substantial."—*Pennsylvania Inquirer.*

"Whatever is said is rendered perfectly intelligible by remarkably well-executed diagrams on steel, leaving nothing for mere vague supposition; and the addition of an introduction to isometrical drawing, linear perspective, and the projection of shadows, winding up with a useful index to technical terms." —*Glasgow Mechanics' Journal.*

☞ The British Government has authorized the use of this book in their schools of art at Somerset House, London, and throughout the kingdom.

Minifie's Geometrical Drawing.

New Edition. Enlarged.

12mo. Cloth. $2.00.

GEOMETRICAL DRAWING. Abridged from the octavo edition, for the use of Schools. Illustrated with 48 steel plates. New edition, enlarged.

"It is well adapted as a text-book of drawing to be used in our High Schools and Academies where this useful branch of the fine arts has been hitherto too much neglected."—*Boston Journal.*

Bell on Iron Smelting.

8vo. Cloth. $6.00.

CHEMICAL PHENOMENA OF IRON SMELTING. An experimental and practical examination of the circumstances which determine the capacity of the Blast Furnace, the Temperature of the Air, and the Proper Condition of the Materials to be operated upon. By I. LOWTHIAN BELL.

"The reactions which take place in every foot of the blast-furnace have been investigated, and the nature of every step in the process, from the introduction of the raw material into the furnace to the production of the pig iron, has been carefully ascertained, and recorded so fully that any one in the trade can readily avail themselves of the knowledge acquired; and we have no hesitation in saying that the judicious application of such knowledge will do much to facilitate the introduction of arrangements which will still further economize fuel, and at the same time permit of the quality of the resulting metal being maintained, if not improved. The volume is one which no practical pig iron manufacturer can afford to be without if he be desirous of entering upon that competition which nowadays is essential to progress, and in issuing such a work Mr. Bell has entitled himself to the best thanks of every member of the trade."—*London Mining Journal.*

King's Notes on Steam.

Thirteenth Edition.

8vo. Cloth. $2.00.

LESSONS AND PRACTICAL NOTES ON STEAM, the Steam-Engine, Propellers, &c., &c., for Young Engineers, Students, and others. By the late W. R. KING, U. S. N. Revised by Chief-Engineer J. W. KING, U. S. Navy.

"This is one of the best, because eminently plain and practical treatises on the Steam Engine ever published.'—*Philadelphia Press.*

This is the thirteenth edition of a valuable work of the late W. H. King, U. S. N. It contains lessons and practical notes on Steam and the Steam Engine, Propellers, etc. It is calculated to be of great use to young marine engineers, students, and others. The text is illustrated and explained by numerous diagrams and representations of machinery.—*Boston Daily Advertiser.*

Text-book at the U. S. Naval Academy, Annapolis.

Burgh's Modern Marine Engineering.

One thick 4to vol. Cloth. $25.00. Half morocco. $30.00.

MODERN MARINE ENGINEERING, applied to Paddle and Screw Propulsion. Consisting of 36 Colored Plates, 259 Practical Wood-cut Illustrations, and 403 pages of Descriptive Matter, the whole being an exposition of the present practice of the following firms: Messrs. J. Penn & Sons; Messrs. Maudslay, Sons & Field; Messrs. James Watt & Co.; Messrs. J. & G. Rennie; Messrs. R. Napier & Sons; Messrs. J. & W. Dudgeon; Messrs. Ravenhill & Hodgson; Messrs. Humphreys & Tenant; Mr. J. T. Spencer, and Messrs. Forrester & Co. By N. P. Burgh, Engineer.

Principal Contents.—General Arrangements of Engines, 11 examples—General Arrangement of Boilers, 14 examples—General Arrangement of Superheaters, 11 examples—Details of Oscillating Paddle Engines, 34 examples—Condensers for Screw Engines, both Injection and Surface, 20 examples—Details of Screw Engines, 20 examples—Cylinders and Details of Screw Engines, 21 examples—Slide Valves and Details, 7 examples—Slide Valve, Link Motion, 7 examples—Expansion Valves and Gear, 10 examples—Details in General, 30 examples—Screw Propeller and Fittings, 13 examples Engine and Boiler Fittings, 28 examples - In relation to the Principles of the Marine Engine and Boiler, 33 examples.

Notices of the Press.

"Every conceivable detail of the Marine Engine, under all its various forms, is profusely, and we must add, admirably illustrated by a multitude of engravings, selected from the best and most modern practice of the first Marine Engineers of the day. The chapter on Condensers is peculiarly valuable. In one word, there is no other work in existence which will bear a moment's comparison with it as an exponent of the skill, talent and practical experience to which is due the splendid reputation enjoyed by many British Marine Engineers."—*Engineer.*

"This very comprehensive work, which was issued in Monthly parts, has just been completed. It contains large and full drawings and copious descriptions of most of the best examples of Modern Marine Engines, and it is a complete theoretical and practical treatise on the subject of Marine Engineering."—*American Artisan.*

This is the only edition of the above work with the beautifully *colored* plates, and it is out of print in England.

Bourne's Treatise on the Steam Engine.

Ninth Edition.

Illustrated. 4to. Cloth. $15.00.

TREATISE ON THE STEAM ENGINE in its various applications to Mines, Mills, Steam Navigation, Railways, and Agriculture, with the theoretical investigations respecting the Motive Power of Heat and the proper Proportions of Steam Engines. Elaborate Tables of the right dimensions of every part, and Practical Instructions for the Manufacture and Management of every species of Engine in actual use. By JOHN BOURNE, being the ninth edition of "A Treatise on the Steam Engine," by the "Artisan Club." Illustrated by thirty-eight plates and five hundred and forty-six wood-cuts.

As Mr. Bourne's work has the great merit of avoiding unsound and immature views, it may safely be consulted by all who are really desirous of acquiring trustworthy information on the subject of which it treats. During the twenty-two years which have elapsed from the issue of the first edition, the improvements introduced in the construction of the steam engine have been both numerous and important, and of these Mr. Bourne has taken care to point out the more prominent, and to furnish the reader with such information as shall enable him readily to judge of their relative value. This edition has been thoroughly modernized, and made to accord with the opinions and practice of the more successful engineers of the present day. All that the book professes to give is given with ability and evident care. The scientific principles which are permanent are admirably explained, and reference is made to many of the more valuable of the recently introduced engines. To express an opinion of the value and utility of such a work as *The Artisan Club's Treatise on the Steam Engine*, which has passed through eight editions already, would be superfluous; but it may be safely stated that the work is worthy the attentive study of all either engaged in the manufacture of steam engines or interested in economizing the use of steam.—*Mining Journal.*

Isherwood's Engineering Precedents.

Two Vols. in One. 8vo. Cloth. $2.50.

ENGINEERING PRECEDENTS FOR STEAM MACHINERY. Arranged in the most practical and useful manner for Engineers. By B. F. ISHERWOOD, Civil Engineer, U. S. Navy. With illustrations.

Ward's Steam for the Million.

New and Revised Edition.

8vo. Cloth. $1.00.

STEAM FOR THE MILLION. A Popular Treatise on Steam and its Application to the Useful Arts, especially to Navigation. By J. H. WARD, Commander U. S. Navy. New and revised edition.

A most excellent work for the young engineer and general reader. Many facts relating to the management of the boiler and engine are set forth with a simplicity of language and perfection of detail that bring the subject home to the reader.—*American Engineer.*

Walker's Screw Propulsion.

8vo. Cloth. 75 cents.

NOTES ON SCREW PROPULSION, its Rise and History. By Capt. W. H. WALKER, U. S. Navy.

Commander Walker's book contains an immense amount of concise practical data, and every item of information recorded fully proves that the various points bearing upon it have been well considered previously to expressing an opinion.—*London Mining Journal.*

Page's Earth's Crust.

18mo. Cloth. 75 cents.

THE EARTH'S CRUST: a Handy Outline of Geology. By DAVID PAGE.

"Such a work as this was much wanted—a work giving in clear and intelligible outline the leading facts of the science, without amplification or irksome details. It is admirable in arrangement, and clear and easy, and, at the same time, forcible in style. It will lead, we hope, to the introduction of Geology into many schools that have neither time nor room for the study of large treatises."—*The Museum.*

Rogers' Geology of Pennsylvania.

3 Vols. 4to, with Portfolio of Maps. Cloth. $30.00.

THE GEOLOGY OF PENNSYLVANIA. A Government Survey. With a general view of the Geology of the United States, Essays on the Coal Formation and its Fossils, and a description of the Coal Fields of North America and Great Britain. By HENRY DARWIN ROGERS, Late State Geologist of Pennsylvania. Splendidly illustrated with Plates and Engravings in the Text.

It certainly should be in every public library throughout the country, and likewise in the possession of all students of Geology. After the final sale of these copies, the work will, of course, become more valuable.

The work for the last five years has been entirely out of the market, but a few copies that remained in the hands of Prof. Rogers, in Scotland, at the time of his death, are now offered to the public, at a price which is even below what it was originally sold for when first published.

Morfit on Pure Fertilizers.

With 28 Illustrative Plates. 8vo. Cloth. $20.00.

A PRACTICAL TREATISE ON PURE FERTILIZERS, and the Chemical Conversion of Rock Guanos, Marlstones, Coprolites, and the Crude Phosphates of Lime and Alumina Generally, into various Valuable Products. By CAMPBELL MORFIT, M.D., F.C.S.

Sweet's Report on Coal.

8vo. Cloth. $3.00.

SPECIAL REPORT ON COAL; showing its Distribution, Classification, and Cost delivered over different routes to various points in the State of New York, and the principal cities on the Atlantic Coast. By S. H. SWEET. With maps.

Colburn's Gas Works of London.

12mo. Boards. 60 cents.

GAS WORKS OF LONDON. By ZERAH COLBURN.

The Useful Metals and their Alloys; Scoffren, Truran, and others.

Fifth Edition.

8vo. Half calf. $3.75.

THE USEFUL METALS AND THEIR ALLOYS, including MINING VENTILATION, MINING JURISPRUDENCE AND METALLURGIC CHEMISTRY employed in the conversion of IRON, COPPER, TIN, ZINC, ANTIMONY, AND LEAD ORES, with their applications to THE INDUSTRIAL ARTS. By John Scoffren, William Truran, William Clay, Robert Oxland, William Fairbairn, W. C. Aitkin, and William Vose Pickett.

Collins' Useful Alloys.

18mo. Flexible. 75 cents.

THE PRIVATE BOOK OF USEFUL ALLOYS and Memoranda for Goldsmiths, Jewellers, etc. By James E. Collins

This little book is compiled from notes made by the Author from the papers of one of the largest and most eminent Manufacturing Goldsmiths and Jewellers in this country, and as the firm is now no longer in existence, and the Author is at present engaged in some other undertaking, he now offers to the public the benefit of his experience, and in so doing he begs to state that all the alloys, etc., given in these pages may be confidently relied on as being thoroughly practicable.

The Memoranda and Receipts throughout this book are also compiled from practice, and will no doubt be found useful to the practical jeweller. —*Shirley, July,* 1871.

Joynson's Metals Used in Construction.

12mo. Cloth. 75 cents.

THE METALS USED IN CONSTRUCTION: Iron, Steel, Bessemer Metal, etc., etc. By Francis Herbert Joynson. Illustrated.

"In the interests of practical science, we are bound to notice this work; and to those who wish further information, we should say, buy it; and the outlay, we honestly believe, will be considered well spent." —*Scientific Review.*

Holley's Ordnance and Armor.

493 Engravings. Half Roan, $10.00. Half Russia, $12.00.

A TREATISE ON ORDNANCE AND ARMOR—Embracing Descriptions, Discussions, and Professional Opinions concerning the MATERIAL, FABRICATION, Requirements, Capabilities, and Endurance of European and American Guns, for Naval, Sea Coast, and Iron-clad Warfare, and their RIFLING, PROJECTILES, and BREECH-LOADING; also, Results of Experiments against Armor, from Official Records, with an Appendix referring to Gun-Cotton, Hooped Guns, etc., etc. By ALEXANDER L. HOLLEY, B. P. 948 pages, 493 Engravings, and 147 Tables of Results, etc.

CONTENTS.

CHAPTER I.—Standard Guns and their Fabrication Described: Section 1. Hooped Guns; Section 2. Solid Wrought Iron Guns; Section 3. Solid Steel Guns; Section 4. Cast-Iron Guns. CHAPTER II.—The Requirements of Guns, Armor: Section 1. The Work to be done; Section 2. Heavy Shot at Low Velocities; Section 3. Small Shot at High Velocities; Section 4. The two Systems Combined; Section 5. Breaching Masonry. CHAPTER III.—The Strains and Structure of Guns: Section 1. Resistance to Elastic Pressure; Section 2. The Effects of Vibration; Section 3. The Effects of Heat. CHAPTER IV.—Cannon Metals and Processes of Fabrication: Section 1. Elasticity and Ductility; Section 2. Cast-Iron; Section 3. Wrought Iron; Section 4. Steel; Section 5. Bronze; Section 6. Other Alloys. CHAPTER V.—Rifling and Projectiles; Standard Forms and Practice Described; Early Experiments; The Centring System; The Compressing System; The Expansion System; Armor Punching Projectiles; Shells for Molten Metal; Competitive Trial of Rifled Guns, 1862; Duty of Rifled Guns: General Uses, Accuracy, Range, Velocity, Strain, Liability of Projectile to Injury; Firing Spherical Shot from Rifled Guns; Material for Armor-Punching Projectiles; Shape of Armor-Punching Projectiles; Capacity and Destructiveness of Shells; Elongated Shot from Smooth Bores; Conclusions; Velocity of Projectiles (Table). CHAPTER VI.—Breech-Loading Advantages and Defects of the System; Rapid Firing and Cooling Guns by Machinery; Standard Breech-Loaders Described. Part Second: Experiments against Armor; Account of Experiments from Official Records in Chronological Order. APPENDIX.—Report on the Application of Gun-Cotton to Warlike Purposes—British Association, 1863; Manufacture and Experiments in England; Guns Hooped with Initial Tension—History; How Guns Burst, by Wiard, Lyman's Accelerating Gun; Endurance of Parrott and Whitworth Guns at Charleston; Hooping old United States Cast-Iron Guns; Endurance and Accuracy of the Armstrong 600-pounder; Competitive Trials with 7-inch Guns.

Peirce's Analytic Mechanics.

4to. Cloth. $10.00.

SYSTEM OF ANALYTIC MECHANICS. Physical and Celestial Mechanics. By BENJAMIN PEIRCE, Perkins Professor of Astronomy and Mathematics in Harvard University, and Consulting Astronomer of the American Ephemeris and Nautical Almanac. Developed in four systems of Analytic Mechanics, Celestial Mechanics, Potential Physics, and Analytic Morphology.

"I have re-examined the memoirs of the great geometers, and have striven to consolidate their latest researches and their most exalted forms of thought into a consistent and uniform treatise. If I have hereby succeeded in opening to the students of my country a readier access to these choice jewels of intellect; if their brilliancy is not impaired in this attempt to reset them; if, in their own constellation, they illustrate each other, and concentrate a stronger light upon the names of their discoverers, and, still more, if any gem which I may have presumed to add is not wholly lustreless in the collection, I shall feel that my work has not been in vain."—*Extract from the Preface.*

Burt's Key to Solar Compass.

Second Edition.

Pocket Book Form. Tuck. $2.50.

KEY TO THE SOLAR COMPASS, and Surveyor's Companion; comprising all the Rules necessary for use in the field; also, Description of the Linear Surveys and Public Land System of the United States, Notes on the Barometer, Suggestions for an outfit for a Survey of four months, etc., etc., etc. By W. A. BURT, U. S. Deputy Surveyor. Second edition.

Chauvenet's Lunar Distances.

8vo. Cloth. $2.00.

NEW METHOD OF CORRECTING LUNAR DISTANCES, and Improved Method of Finding the Error and Rate of a Chronometer, by equal altitudes. By WM. CHAUVENET, LL.D., Chancellor of Washington University of St. Louis.

Jeffers' Nautical Surveying.

Illustrated with 9 Copperplates and 31 Wood-cut Illustrations. 8vo. Cloth. $5.00.

NAUTICAL SURVEYING. By WILLIAM N. JEFFERS, Captain U. S. Navy.

Many books have been written on each of the subjects treated of in the sixteen chapters of this work; and, to obtain a complete knowledge of geodetic surveying requires a profound study of the whole range of mathematical and physical sciences; but a year of preparation should render any intelligent officer competent to conduct a nautical survey.

CONTENTS.—Chapter I. Formulæ and Constants Useful in Surveying II. Distinctive Character of Surveys. III. Hydrographic Surveying under Sail; or, Running Survey. IV. Hydrographic Surveying of Boats; or, Harbor Survey. V. Tides—Definition of Tidal Phenomena—Tidal Observations. VI. Measurement of Bases—Appropriate and Direct. VII. Measurement of the Angles of Triangles—Azimuths—Astronomical Bearings. VIII. Corrections to be Applied to the Observed Angles. IX. Levelling—Difference of Level. X. Computation of the Sides of the Triangulation—The Three-point Problem. XI. Determination of the Geodetic Latitudes, Longitudes, and Azimuths, of Points of a Triangulation. XII. Summary of Subjects treated of in preceding Chapters—Examples of Computation by various Formulæ. XIII. Projection of Charts and Plans. XIV. Astronomical Determination of Latitude and Longitude. XV. Magnetic Observations. XVI. Deep Sea Soundings. XVII. Tables for Ascertaining Distances at Sea, and a full Index.

List of Plates.

Plate I. Diagram Illustrative of the Triangulation. II. Specimen Page of Field Book. III. Running Survey of a Coast. IV. Example of a Running Survey from Belcher. V. Flying Survey of an Island. VI. Survey of a Shoal. VII. Boat Survey of a River. VIII. Three-Point Problem. IX. Triangulation.

Coffin's Navigation.

Fifth Edition.

12mo. Cloth. $3.50.

NAVIGATION AND NAUTICAL ASTRONOMY. Prepared for the use of the U. S. Naval Academy. By J. H. C. COFFIN, Prof. of Astronomy, Navigation and Surveying, with 52 woodcut illustrations.

Clark's Theoretical Navigation.

8vo. Cloth. $3.00.

THEORETICAL NAVIGATION AND NAUTICAL ASTRONOMY. By LEWIS CLARK, Lieut.-Commander, U. S. Navy. Illustrated with 41 Wood-cuts, including the Vernier.

Prepared for Use at the U. S. Naval Academy.

The Plane Table.

Illustrated. 8vo. Cloth. $2.00.

ITS USES IN TOPOGRAPHICAL SURVEYING. From the Papers of the U. S. Coast Survey.

This work gives a description of the Plane Table employed at the U. S. Coast Survey Office, and the manner of using it.

Pook on Shipbuilding.

8vo. Cloth. $5.00.

METHOD OF COMPARING THE LINES AND DRAUGHTING VESSELS PROPELLED BY SAIL OR STEAM, including a Chapter on Laying off on the Mould-Loft Floor. By SAMUEL M. POOK, Naval Constructor. 1 vol., 8vo. With illustrations. Cloth. $5.00.

Brunnow's Spherical Astronomy.

8vo. Cloth. $6.50.

SPHERICAL ASTRONOMY. By F. BRUNNOW, Ph. Dr. Translated by the Author from the Second German edition.

Van Buren's Formulas.

8vo. Cloth. $2.00.

INVESTIGATIONS OF FORMULAS, for the Strength of the Iron Parts of Steam Machinery. By J. D. VAN BUREN, Jr., C. E. Illustrated.

This is an analytical discussion of the formulæ employed by mechanical engineers in determining the rupturing or crippling pressure in the different parts of a machine. The formulæ are founded upon the principle, that the different parts of a machine should be equally strong, and are developed in reference to the ultimate strength of the material in order to leave the choice of a factor of safety to the judgment of the designer.—*Silliman's Journal.*

Joynson on Machine Gearing.

8vo. Cloth. $2.00.

THE MECHANIC'S AND STUDENT'S GUIDE in the Designing and Construction of General Machine Gearing, as Eccentrics, Screws, Toothed Wheels, etc., and the Drawing of Rectilineal and Curved Surfaces ; with Practical Rules and Details. Edited by FRANCIS HERBERT JOYNSON. Illustrated with 18 folded plates.

"The aim of this work is to be a guide to mechanics in the designing and construction of general machine-gearing. This design it well fulfils, being plainly and sensibly written, and profusely illustrated."—*Sunday Times.*

Barnard's Report, Paris Exposition, 1867.

Illustrated. 8vo. Cloth. $5.00.

REPORT ON MACHINERY AND PROCESSES ON THE INDUSTRIAL ARTS AND APPARATUS OF THE EXACT SCIENCES. By F. A. P. BARNARD, LL.D.—Paris Universal Exposition, 1867.

"We have in this volume the results of Dr. Barnard's study of the Paris Exposition of 1867, in the form of an official Report of the Government. It is the most exhaustive treatise upon modern inventions that has appeared since the Universal Exhibition of 1851, and we doubt if anything equal to it has appeared this century."—*Journal Applied Chemistry.*

Engineering Facts and Figures.

18mo. Cloth. $2.50 per Volume.

AN ANNUAL REGISTER OF PROGRESS IN MECHANICAL ENGINEERING AND CONSTRUCTION, for the Years 1863–64–65–66–67–68. Fully illustrated. 6 volumes.

Each volume sold separately.

Beckwith's Pottery.

8vo. Paper. 60 cents.

OBSERVATIONS ON THE MATERIALS and Manufacture of Terra-Cotta, Stone-Ware, Fire-Brick, Porcelain and Encaustic Tiles, with Remarks on the Products exhibited at the London International Exhibition, 1871. By ARTHUR BECKWITH, Civil Engineer.

"Everything is noticed in this book which comes under the head of Pottery, from fine porcelain to ordinary brick, and aside from the interest which all take in such manufactures, the work will be of considerable value to followers of the ceramic art."—*Evening Mail.*

Dodd's Dictionary of Manufactures, etc.

12mo. Cloth. $2.00.

DICTIONARY OF MANUFACTURES, MINING, MACHINERY, AND THE INDUSTRIAL ARTS. By GEORGE DODD.

This work, a small book on a great subject, treats, in alphabetical arrangement, of those numerous matters which come generally within the range of manufactures and the productive arts. The raw materials—animal, vegetable, and mineral—whence the manufactured products are derived, are succinctly noticed in connection with the processes which they undergo, but not as subjects of natural history. The operations of the Mine and the Mill, the Foundry and the Forge, the Factory and the Workshop, are passed under review. The principal machines and engines, tools and apparatus, concerned in manufacturing processes, are briefly described. The scale on which our chief branches of national industry are conducted, in regard to values and quantities, is indicated in various ways.

Stuart's Civil and Military Engineering of America.

8vo. Illustrated. Cloth. $5.00.

THE CIVIL AND MILITARY ENGINEERS OF AMERICA. By General CHARLES B. STUART, Author of "Naval Dry Docks of the United States," etc., etc. Embellished with nine finely executed portraits on steel of eminent engineers, and illustrated by engravings of some of the most important and original works constructed in America.

Containing sketches of the Life and Works of Major Andrew Ellicott, James Geddes (with Portrait), Benjamin Wright (with Portrait), Canvass White (with Portrait), David Stanhope Bates, Nathan S. Roberts, Gridley Bryant (with Portrait), General Joseph G. Swift, Jesse L. Williams (with Portrait), Colonel William McRee, Samuel H. Kneass, Captain John Childe (with Portrait), Frederick Harbach, Major David Bates Douglas (with Portrait), Jonathan Knight, Benjamin H. Latrobe (with Portrait), Colonel Charles Ellet, Jr. (with Portrait), Samuel Forrer, William Stuart Watson, John A. Roebling.

Alexander's Dictionary of Weights and Measures.

8vo. Cloth. $3.50.

UNIVERSAL DICTIONARY OF WEIGHTS AND MEASURES, Ancient and Modern, reduced to the standards of the United States of America. By J. H. ALEXANDER. New edition. 1 vol.

"As a standard work of reference, this book should be in every library; it is one which we have long wanted, and it will save much trouble and research."—*Scientific American.*

Gouge on Ventilation.

Third Edition Enlarged.

8vo. Cloth. $2.00.

NEW SYSTEM OF VENTILATION, which has been thoroughly tested under the patronage of many distinguished persons. By HENRY A. GOUGE, with many illustrations.

Saeltzer's Acoustics.

12mo. Cloth. $2.00.

TREATISE ON ACOUSTICS in Connection with Ventilation. With a new theory based on an important discovery, of facilitating clear and intelligible sound in any building. By ALEXANDER SAELTZER.

"A practical and very sound treatise on a subject of great importance to architects, and one to which there has hitherto been entirly too little attention paid. The author's theory is, that, by bestowing proper care upon the point of Acoustics, the requisite ventilation will be obtained, and *vice versa.*—*Brooklyn Union.*

Myer's Manual of Signals.

New Edition. Enlarged.

12mo. 48 Plates full Roan. $5.00.

MANUAL OF SIGNALS, for the Use of Signal Officers in the Field, and for Military and Naval Students, Military Schools, etc. A new edition, enlarged and illustrated. By Brig.-Gen. ALBERT J. MYER, Chief Signal Officer of the Army, Colonel of the Signal Corps during the War of the Rebellion.

Larrabee's Secret Letter and Telegraph Code.

18mo. Cloth. $1.00.

CIPHER AND SECRET LETTER AND TELEGRAPHIC CODE, with Hogg's Improvements. The most perfect secret Code ever invented or discovered. Impossible to read without the Key. Invaluable for Secret, Military, Naval, and Diplomatic Service, as well as for Brokers, Bankers, and Merchants. By C. S. LARRABEE, the original inventor of the scheme.

Hunt's Designs for Central Park Gateways.

4to. Cloth. $5.00.

DESIGNS FOR THE GATEWAYS OF THE SOUTHERN ENTRANCES TO THE CENTRAL PARK. By RICHARD M. HUNT. With a description of the designs.

Pickert and Metcalf's Art of Graining.

1 vol. 4to. Cloth. $10.00.

THE ART OF GRAINING. How Acquired and How Produced, with description of colors and their application. By CHARLES PICKERT and ABRAHAM METCALF. Beautifully illustrated with 42 tinted plates of the various woods used in interior finishing. Tinted paper.

The authors present here the result of long experience in the practice of this decorative art, and feel confident that they hereby offer to their brother artisans a reliable guide to improvement in the practice of graining.

Portrait Gallery of the War.

60 fine Portraits on Steel. Royal 8vo. Cloth. $6.00.

PORTRAIT GALLERY OF THE WAR, CIVIL, MILITARY AND NAVAL. A Biographical Record. Edited by FRANK MOORE.

One Law in Nature.

12mo. Cloth. $1.50.

ONE LAW IN NATURE. By Capt. H. M. LAZELLE, U. S. A. A New Corpuscular Theory, comprehending Unity of Force, Identity of Matter, and its Multiple Atom Constitution, applied to the Physical Affections or Modes of Energy.

Ernst's Manual of Military Engineering.

193 Wood Cuts and 3 Lithographed Plates. 12mo. Cloth. $5.00.

A MANUAL OF PRACTICAL MILITARY ENGINEERING. Prepared for the use of the Cadets of the U. S. Military Academy, and for Engineer Troops. By Capt. O. H. Ernst, Corps of Engineers, Instructor in Practical Military Engineering, U. S. Military Academy.

Church's Metallurgical Journey.

24 Illustrations. 8vo. Cloth. $2.00.

NOTES OF A METALLURGICAL JOURNEY IN EUROPE. By John A. Church, Engineer of Mines.

Blake's Precious Metals.

8vo. Cloth. $2.00.

REPORT UPON THE PRECIOUS METALS: Being Statistical Notices of the principal Gold and Silver producing regions of the World. Represented at the Paris Universal Exposition. By William P. Blake, Commissioner from the State of California.

Clevenger's Surveying.

Illustrated Pocket Form. Morocco Gilt. $2.50.

A TREATISE ON THE METHOD OF GOVERNMENT SURVEYING, as prescribed by the United States Congress. and Commissioner of the General Land Office. With complete Mathematical, Astronomical and Practical Instructions, for the use of the United States Surveyors in the Field, and Students who contemplate engaging in the business of Public Land Surveying. By S. R. Clevenger, U. S. Deputy Surveyor.

"The reputation of the author as a surveyor guarantees an exhaustive treatise on this subject."—*Dakota Register.*

"Surveyors have long needed a text-book of this description.—*The Press.*

Bow on Bracing.

156 Illustrations on Stone. 8vo. Cloth. $1.50.

A TREATISE ON BRACING, with its application to Bridges and other Structures of Wood or Iron. By ROBERT HENRY BOW, C. E.

Howard's Earthwork Mensuration.

8vo. Illustrated. Cloth. $1.50.

EARTHWORK MENSURATION ON THE BASIS OF THE PRISMOIDAL FORMULÆ. Containing simple and labor-saving method of obtaining Prismoidal Contents directly from End Areas. Illustrated by Examples, and accompanied by Plain Rules for practical uses. By CONWAY R. HOWARD, Civil Engineer, Richmond, Va.

McAlpine's Modern Engineering.

Second Edition. 8vo. Cloth. $1.50.

MODERN ENGINEERING. A Lecture delivered at the American Institute in New York. By WILLIAM J. MCALPINE.

Mowbray's Tri-Nitro-Glycerine.

8vo. Cloth. Illustrated. $3.00.

TRI-NITRO-GLYCERINE, as applied in the Hoosac Tunnel, and to Submarine Blasting, Torpedoes, Quarrying, etc. Being the result of six years' observation and practice during the manufacture of five hundred thousand pounds of this explosive, Mica Blasting Powder, Dynamites; with an account of the various Systems of Blasting by Electricity, Priming Compounds, Explosives, etc., etc. By GEORGE M. MOWBRAY, Operative Chemist, with thirteen illustrations, tables, and appendix. Third Edition. Re-written.

Wanklyn's Milk Analysis.

12mo. Cloth. $1.00.

MILK ANALYSIS. A Practical Treatise on the Examination of Milk, and its Derivatives, Cream, Butter and Cheese. By J. ALFRED WANKLYN, M. R. C. S.

Toner's Dictionary of Elevations.

8vo. Paper, $3.00. Cloth, $3.75.

DICTIONARY OF ELEVATIONS AND CLIMATIC REGISTER OF THE UNITED STATES. Containing, in addition to Elevations, the Latitude, Mean Annual Temperature, and the total Annual Rain Fall of many localities; with a brief Introduction on the Orographic and Physical Peculiarities of North America. By J. M. TONER, M. D.

Adams. Sewers and Drains.

(*In Press.*)

SEWERS AND DRAINS FOR POPULOUS DISTRICTS. Embracing Rules and Formulas for the dimensions of Sanitary Engineers. By JULIUS W. ADAMS, Chief Engineer of the Board of City Works, Brooklyn.

Prescott's Proximate Organic Analysis.

12mo. Cloth. $1.75.

OUTLINES OF PROXIMATE ORGANIC ANALYSIS for the Identification, Separation, and Quantitative Determination of the more commonly occurring Organic Compounds. By ALBERT B. PRESCOTT, Professor of Organic and Applied Chemistry in the University of Michigan.

Prescott's Alcoholic Liquors.

12mo. Cloth. $1.50.

CHEMICAL EXAMINATION OF ALCOHOLIC LIQUORS. A Manual of the Constituents of the Distilled Spirits and Fermented Liquors of Commerce, and their Qualitative and Quantitative Determinations. By ALBERT B. PRESCOTT, Professor of Organic and Applied Chemistry in the University of Michigan.

Greene's Bridge Trusses.

8vo. Illustrated. Cloth. $2.00.

GRAPHICAL METHOD FOR THE ANALYSIS OF BRIDGE TRUSSES, extended to Continuous Girders and Draw Spans. By CHARLES E. GREENE, A.M., Professor of Civil Engineering, University of Michigan. Illustrated by three folding plates.

Butler's Projectiles and Rifled Cannon.

4to. 32 Plates. Cloth. In press.

PROJECTILES AND RIFLED CANNON. A Critical Discussion of the Principal Systems of Rifling and Projectiles, with Practical Suggestions for their Improvement, as embraced in a Report to the Chief of Ordnance, U.S.A. By Capt. JOHN S. BUTLER, Ordnance Corps, U.S.A.

Van Nostrand's Science Series.

It is the intention of the Publisher of this Series to issue them at intervals of about a month. They will be put up in a uniform, neat and attractive form, 18mo, fancy boards. The subjects will be of an eminently scientific character, and embrace as wide a range of topics as possible, all of the highest character.

Price, 50 Cents Each.

1.

CHIMNEYS FOR FURNACES, FIRE-PLACES, AND STEAM BOILERS. By R. ARMSTRONG, C. E.

2.

STEAM BOILER EXPLOSIONS. By ZERAH COLBURN.

3.

PRACTICAL DESIGNING OF RETAINING WALLS By ARTHUR JACOB, A. B. With Illustrations.

4.

PROPORTIONS OF PINS USED IN BRIDGES. By CHARLES E. BENDER, C. E. With Illustrations.

5.

VENTILATION OF BUILDINGS. By W. F. BUTLER. With Illustrations.

6.

ON THE DESIGNING AND CONSTRUCTION OF STORAGE RESERVOIRS. By ARTHUR JACOB. With Illustrations.

7.

SURCHARGED AND DIFFERENT FORMS OF RETAINING WALLS. By JAMES S. TATE, C. E.

8.

A TREATISE ON THE COMPOUND ENGINE. By JOHN TURNBULL. With Illustrations.

9.

FUEL. By C. W. SIEMENS to which is appended the Value of Artificial Fuels as compared with Coal. By J. WORMALD, C. E.

*** Other works in preparation.

10.

COMPOUND ENGINES. Translated from the French of A. MALLET. Illustrated.

11.

THEORY OF ARCHES. By Prof. W. ALLAN, of the Washington and Lee College. Illustrated.

12.

A PRACTICAL THEORY OF VOUSSOIR ARCHES. By WILLIAM CAIN, C.E. Illustrated.

13.

A PRACTICAL TREATISE ON THE GASES MET WITH IN COAL-MINES. By the late J. J. ATKINSON, Government Inspector of Mines for the County of Durham, England.

14.

FRICTION OF AIR IN MINES. By J. J. ATKINSON, Author of "A Practical Treatise on the Gases met with in Coal-Mines."

15.

SKEW ARCHES. By Prof. E. W. HYDE, C.E. Illustrated with numerous engravings and three folded plates.

SILVER MINING REGIONS OF COLORADO, with some account of the different Processes now being introduced for working the Gold Ores of that Territory. By J. P. WHITNEY. 12mo. Paper. 25 cents.

COLORADO: SCHEDULE OF ORES contributed by sundry persons to the Paris Universal Exposition of 1867, with some information about the Region and its Resources. By J. P. WHITNEY, Commissioner from the Territory. 8vo. Paper, with Maps. 25 cents.

THE SILVER DISTRICTS OF NEVADA. With Map. 8vo. Paper. 35 cents.

ARIZONA: ITS RESOURCES AND PROSPECTS. By Hon. R. C. McCORMICK, Secretary of the Territory. With Map. 8vo. Paper. 25 cents.

MONTANA AS IT IS. Being a general description of its Resources, both Mineral and Agricultural; including a complete description of the face of the country, its climate, etc. Illustrated with a Map of the Territory, showing the different Roads and the location of the different Mining Districts. To which is appended a complete Dictionary of THE SNAKE LANGUAGE, and also of the famous Chinnook Jargon, with numerous critical and explanatory Notes. By GRANVILLE STUART. 8vo. Paper. $2.00.

RAILWAY GAUGES. A Review of the Theory of Narrow Gauges as applied to Main Trunk Lines of Railway. By SILAS SEYMOUR, Genl. Consulting Engineer. 8vo. Paper. 50 cents.

REPORT made to the President and Executive Board of the Texas Pacific Railroad. By Gen. G. P. BUELL, Chief Engineer. 8vo. Paper. 75 cents.

www.ingramcontent.com/pod-product-compliance
Lightning Source LLC
LaVergne TN
LVHW012326100826
845148LV00017B/351